W9-AOZ-136

Plasma Science

From Fundamental Research to Technological Applications

Panel on Opportunities in Plasma Science and Technology

Plasma Science Committee
Board on Physics and Astronomy
Commission on Physical Sciences, Mathematics, and Applications
National Research Council

NATIONAL ACADEMY PRESS
Washington, D.C. 1995

NOTICE: The project that is the subject of this report was approved by the Governing Board of the National Research Council, whose members are drawn from the councils of the National Academy of Sciences, the National Academy of Engineering, and the Institute of Medicine. The members of the committee responsible for the report were chosen for their special competences and with regard for appropriate balance.

This report has been reviewed by a group other than the authors according to procedures approved by a Report Review Committee consisting of members of the National Academy of Sciences, the National Academy of Engineering, and the Institute of Medicine.

This project was supported by the Department of Energy under Contract No. DE-FG05-88ER53279, the National Science Foundation under Grant No. PHY-9100105, and the Office of Naval Research under Contract No. N00014-J-1728.

Library of Congress Catalog Card No. 94-69693
International Standard Book No. 0-309-05231-9

Cover: A snapshot of the electron density distribution in a magnetized, pure-electron plasma. These plasmas are nearly ideal, inviscid, two-dimensional fluids and are being used to study the relaxation and self-organization of fluid turbulence (see Plate 2 for details). (Courtesy of C.F. Driscoll, University of California, San Diego.)

Additional copies of this report are available from:

National Academy Press
2101 Constitution Avenue, NW
Box 285
Washington, DC 20055
800-624-6242
202-334-3313 (in the Washington Metropolitan Area)

Copyright 1995 by the National Academy of Sciences. All rights reserved.

Printed in the United States of America

PANEL ON OPPORTUNITIES IN PLASMA SCIENCE AND TECHNOLOGY

CLIFFORD SURKO, University of California, San Diego, *Co-Chair*
JOHN AHEARNE, Sigma Xi, The Scientific Research Society, *Co-Chair*
PETER BANKS, University of Michigan
THOMAS BIRMINGHAM, NASA Goddard Space Flight Center
MICHAEL BOYLE, Bondtronix, Inc.
RONALD C. DAVIDSON, Princeton University
JONAH JACOB, Science Research Laboratory, Inc.
MIKLOS PORKOLAB, Massachusetts Institute of Technology
EDWIN SALPETER, Cornell University
ROBERTA SAXON, SRI International
SAM TREIMAN, Princeton University
HERBERT YORK, University of California, San Diego (retired)
ELLEN ZWEIBEL, University of Colorado

RONALD D. TAYLOR, Senior Program Officer (1992–1994)
DANIEL F. MORGAN, Program Officer

PLASMA SCIENCE COMMITTEE

RAVI SUDAN, Cornell University, *Chair*
RICHARD A. GOTTSCHO, AT&T Bell Laboratories, *Vice Chair*
STEVEN C. COWLEY, University of California, Los Angeles
JAMES DAKIN, GE Lighting
ROY GOULD, California Institute of Technology
RICHARD D. HAZELTINE, University of Texas at Austin
MARY KATHERINE HUDSON, Dartmouth College
WILLIAM L. KRUER, Lawrence Livermore National Laboratory
MICHAEL LIEBERMAN, University of California, Berkeley
CHUAN S. LIU, University of Maryland
NATHAN RYNN, University of California, Irvine
ELLEN ZWEIBEL, University of Colorado

Former Members of the Committee Who Were Active During the Period of the Study

JONATHAN ARONS, University of California, Berkeley
MAHA ASHOUR-ABDALLA, University of California, Los Angeles
IRA BERNSTEIN, Yale University
E.M. CAMPBELL, Lawrence Livermore National Laboratory
RONALD C. DAVIDSON, Princeton University
ALAN GARSCADDEN, Wright Research and Development Center
ROBERT L. McCRORY, JR., University of Rochester
FRANCIS W. PERKINS, Princeton University
JOSEPH PROUD, GTE Laboratories Incorporated
NORMAN ROSTOKER, University of California, Irvine

RONALD D. TAYLOR, Senior Program Officer (1992–1994)
DANIEL F. MORGAN, Program Officer

BOARD ON PHYSICS AND ASTRONOMY

DAVID N. SCHRAMM, University of Chicago, *Chair*
ROBERT C. DYNES, University of California, San Diego, *Vice Chair*
LLOYD ARMSTRONG, JR., University of Southern California
DAVID H. AUSTON, Rice University
DAVID E. BALDWIN, Lawrence Livermore National Laboratory
PRAVEEN CHAUDHARI, IBM T.J. Watson Research Center
FRANK DRAKE, University of California, Santa Cruz
HANS FRAUENFELDER, Los Alamos National Laboratory
JEROME I. FRIEDMAN, Massachusetts Institute of Technology
MARGARET J. GELLER, Harvard-Smithsonian Center for Astrophysics
MARTHA P. HAYNES, Cornell University
WILLIAM KLEMPERER, Harvard University
AL NARATH, Sandia National Laboratories
JOSEPH M. PROUD, GTE Corporation (retired)
ROBERT C. RICHARDSON, Cornell University
JOHANNA STACHEL, State University of New York at Stony Brook
DAVID WILKINSON, Princeton University
SIDNEY WOLFF, National Optical Astronomy Observatories

DONALD C. SHAPERO, Director
ROBERT L. RIEMER, Associate Director
DANIEL F. MORGAN, Program Officer
NATASHA CASEY, Senior Administrative Associate
STEPHANIE Y. SMITH, Project Assistant

COMMISSION ON PHYSICAL SCIENCES, MATHEMATICS, AND APPLICATIONS

RICHARD N. ZARE, Stanford University, *Chair*
RICHARD S. NICHOLSON, American Association for the Advancement of Science, *Vice Chair*
STEPHEN L. ADLER, Institute for Advanced Study, Princeton
SYLVIA T. CEYER, Massachusetts Institute of Technology
SUSAN L. GRAHAM, University of California, Berkeley
ROBERT J. HERMANN, United Technologies Corporation
RHONDA J. HUGHES, Bryn Mawr College
SHIRLEY A. JACKSON, Rutgers University
KENNETH I. KELLERMANN, National Radio Astronomy Observatory
HANS MARK, University of Texas at Austin
THOMAS A. PRINCE, California Institute of Technology
JEROME SACKS, National Institute of Statistical Sciences
L.E. SCRIVEN, University of Minnesota
LEON K. SILVER, California Institute of Technology
CHARLES P. SLICHTER, University of Illinois at Urbana-Champaign
ALVIN W. TRIVELPIECE, Oak Ridge National Laboratory
SHMUEL WINOGRAD, IBM T.J. Watson Research Center
CHARLES A. ZRAKET, Mitre Corporation (retired)

NORMAN METZGER, Executive Director

Preface

In the mid-1980s, the plasma physics volume of the series *Physics Through the 1990s* (National Research Council, National Academy Press, Washington, D.C., 1986) signaled problems for plasma science in the United States, particularly with regard to the basic aspects of the science. In the years that followed, there developed a widespread feeling in the plasma science community that something systematic needed to be done to address these issues. Out of this concern, the Plasma Science Committee of the Board on Physics and Astronomy was created in 1988. Following its establishment, plans were begun to undertake this study. With funding from the National Science Foundation, the Department of Energy, and the Office of Naval Research, the Panel on Opportunities in Plasma Science and Technology was appointed in May 1992, and the study began.

Approximately half of the 13-member panel consisted of experts in the many facets of plasma science considered in this report and half of scientists outside the field, with one of the co-chairs selected as a person with experience in science policy. Three of the members are from industry; one is from a government laboratory and one from an independent research society; and the remaining eight are from academe.

The task statement to the panel requested that this study examine virtually all aspects of plasma science and technology in the United States, assess the health of basic plasma science as a research enterprise, and identify and address key issues in the field. Specifically, the panel was charged with the task of conducting an assessment of plasma science that included beams, accelerators, and coherent radiation sources; single-species plasmas and atomic traps; basic plasma science in magnetic confinement and inertial fusion devices; space plasma physics; astrophysics; low-temperature plasmas; and theoretical and computational plasma science. It was directed to address the following:

1. Assess the health of basic plasma science in the United States as a research enterprise: (a) Identify and describe selected scientific opportunities. (b) Identify and describe selected technological opportunities. (c) Assess and prioritize new opportunities for research using the criteria of intellectual challenge, prospects for illumination of classic research questions, connection with other fields of science, and potential for applications. (d) Assess applications using the criteria of potential for contributing to industrial competitiveness, national defense, human health, and other aspects of human welfare.

2. Identify and address the issues in the field, including the following: (a) Evaluate the quality and size of the educational programs in plasma science in light of the nation's future needs. (b) Assess the institutional infrastructure in which plasma science is conducted, and identify changes that would improve the research and educational effort. (c) Characterize the basic experimental facilities needed to increase scientific productivity. (d) Develop a research strategy that is responsive to the issues. (e) Compare the U.S. program with those of Japan and Western Europe, and identify opportunities for international cooperation. (f) Identify the interactions and synergism with other areas of physics, chemistry, mathematics, and astronomy. (g) Assess the linkage of theory and experiment. (h) Assess manpower requirements and the prospects for meeting them. (i) Identify the users of plasma science and their needs.

3. Make recommendations to federal agencies and to the community that address these issues.

During the course of the study, the panel held three two-day meetings and two lengthy teleconferences. As part of the process, the panel took steps to solicit input from the plasma science community. Letters were sent to 200 scientists and engineers, requesting their input on the issues raised in the charge to the panel. This list was selected from the list of Fellows of the Plasma Physics Division of the American Physical Society (90), and it also included others suggested by members of the panel (65) and by grant officers involved in funding plasma science (45). The letters went to university faculty and staff (90), industrial scientists (25), staff at national laboratories (50), and others (5). A separate, more specialized survey was sent to 33 experimentalists engaged in basic plasma physics research. Input was also solicited by announcements of the panel's work that appeared in the newsletters of the American Geophysical Union, the American Physical Society, the Plasma Physics Division of the American Physical Society, the Committee on Plasma Science of the Institute of Electrical and Electronics Engineers (IEEE), and the University Fusion Associates. Town meetings were held at American Physical Society Plasma Physics Division meetings and the Gaseous Electronics Conference. There is general agreement from these sources on the themes expressed in this report: There is concern about the decline in basic plasma science, particularly in the area of basic plasma experimentation and other small-scale research efforts, and basic plasma science is perceived to lack a "home" in the federal agencies.

Also during the course of the study, the panel heard presentations from grant officers involved in funding plasma science from the Air Force Office of Scientific Research, the Advanced Research Projects Agency, the Department of Energy, the National Aeronautics and Space Administration, the National Science Foundation, and the Office of Naval Research.

The task statement requested that the panel assess specific areas of plasma science, such as beams, accelerators, and coherent radiation sources (called *topical areas* in the report), and *broad areas of plasma science*, including fundamental plasma experiments, theoretical and computational plasma physics, and education in plasma science. At the first meeting of the panel, these areas were renamed slightly and the topical area of low-temperature plasmas was added, since it had been omitted from the task statement through an oversight. The resulting seven topical areas are assessed in Part II of the report, and the three broad areas of plasma science are assessed in Part III. Part IV consists of some concluding remarks.

During the course of the study, the panel had numerous discussions about the desirability of establishing organizational units specifically devoted to plasma science in the relevant federal agencies. Many members of the plasma science community who were consulted strongly advocated the establishment of such homes, believing that they are needed if basic plasma science is to be given the focused attention and increased support that the panel recommends. While this subject is beyond the scope of the panel's work, the panel suggests that the federal government might give this issue further consideration.

The National Academy of Sciences is a private, nonprofit, self-perpetuating society of distinguished scholars engaged in scientific and engineering research, dedicated to the furtherance of science and technology and to their use for the general welfare. Upon the authority of the charter granted to it by Congress in 1863, the Academy has a mandate that requires it to advise the federal government on scientific and technical matters. Dr. Bruce Alberts is president of the National Academy of Sciences.

The National Academy of Engineering was established in 1964, under the charter of the National Academy of Sciences, as a parallel organization of outstanding engineers. It is autonomous in its administration and in the selection of its members, sharing with the National Academy of Sciences the responsibility for advising the federal government. The National Academy of Engineering also sponsors engineering programs aimed at meeting national needs, encourages education and research, and recognizes the superior achievements of engineers. Dr. Robert M. White is president of the National Academy of Engineering.

The Institute of Medicine was established in 1970 by the National Academy of Sciences to secure the services of eminent members of appropriate professions in the examination of policy matters pertaining to the health of the public. The Institute acts under the responsibility given to the National Academy of Sciences by its congressional charter to be an advisor to the federal government and, upon its own initiative, to identify issues of medical care, research, and education. Dr. Kenneth I. Shine is president of the Institute of Medicine.

The National Research Council was established by the National Academy of Sciences in 1916 to associate the broad community of science and technology with the Academy's purposes of furthering knowledge and of advising the federal government. Functioning in accordance with general policies determined by the Academy, the Council has become the principal operating agency of both the National Academy of Sciences and the National Academy of Engineering in providing services to the government, the public, and the scientific and engineering communities. The Council is administered jointly by both Academies and the Institute of Medicine. Dr. Bruce Alberts and Dr. Robert M. White are chairman and vice chairman, respectively, of the National Research Council.

Acknowledgments

In preparing this report, the Panel on Opportunities in Plasma Science and Technology has benefited greatly from the assistance of many members of the plasma science community. We are particularly indebted to the former chairs of the Plasma Science Committee of the Board on Physics and Astronomy, C.F. Kennel and F.W. Perkins, and the present chair, Ravi Sudan, for their advice and help. The other members of the Plasma Science Committee also provided valuable advice during the course of the study.

The panel would like to acknowledge the following colleagues for the extensive advice and assistance they provided in assembling the broad range of material covered in this report and for critical reading of various portions of it: Jonathan Arons, University of California, Berkeley; Ira B. Bernstein, Yale University; John Bollinger, National Institute of Standards and Technology, Boulder, Colorado; Keith H. Burrell, GA Technologies, Inc.; Vincent S. Chan, GA Technologies, Inc.; Xing Chen, Science Research Laboratory, Inc.; Samuel A. Cohen, Princeton Plasma Physics Laboratory; Bruce Danly, Plasma Fusion Center, Massachusetts Institute of Technology; Luiz Da Silva, Lawrence Livermore National Laboratory; Patrick Diamond, University of California, San Diego; Paul Drake, Lawrence Livermore National Laboratory; C. Fred Driscoll, University of California, San Diego; Eduardo Epperlein, University of Rochester Laboratory for Laser Energetics; Joel Fajans, University of California, Berkeley; Walter Gekelman, University of California, Los Angeles; Brian Gilchrist, University of Michigan; Martin Goldman, University of Colorado; Tamas I. Gombosi, University of Michigan; Daniel Goodman, Science Research Laboratory, Inc.; Richard A. Gottscho, AT&T Bell Laboratories; Roy W. Gould, California Institute of Technology; Hans Griem, University of Maryland; Larry R. Grisham, Princeton

Plasma Physics Laboratory; Richard Hazeltine, University of Texas; Noah Hershkowitz, University of Wisconsin; Chuck Hooper, University of Florida; Mary Hudson, Dartmouth College; Chandrashekhar Joshi, University of California, Los Angeles; Robert Kessler, Textron Defense Systems; William Kruer, Lawrence Livermore National Laboratory; Stephen Lane, Lawrence Livermore National Laboratory; Richard Lee, Lawrence Livermore National Laboratory; Bruce Lipschultz, Plasma Fusion Center, Massachusetts Institute of Technology; James F. Lyon, Oak Ridge National Laboratory; James Maggs, University of California, Los Angeles; Earl S. Marmar, Plasma Fusion Center, Massachusetts Institute of Technology; Dennis Mathews, Lawrence Livermore National Laboratory; Jakob Maya, Matsushita Electrical Works, R&D Laboratory; Kevin M. McGuire, Princeton Plasma Physics Laboratory; George Morales, University of California, Los Angeles; Andrew Nagy, University of Michigan; Torsten Neubert, University of Michigan; Francis W. Perkins, Princeton Plasma Physics Laboratory; Arthur V. Phelps, JILA, University of Colorado (retired); Stewart C. Prager, University of Wisconsin; Juan Ramirez, Sandia National Laboratories; Barrett Ripin, American Physical Society; Gerald L. Rogoff, Sylvania, Inc.; Louis Rosocha, Los Alamos National Laboratory; Norman Rostoker, University of California, Los Angeles; Andrew Schmitt, Naval Research Laboratory; Wolf Seka, University of Rochester Laboratory for Laser Energetics; Gary Selwyn, Los Alamos National Laboratory; Frederick Skiff, University of Maryland; Reiner Stenzel, University of California, Los Angeles; Raul Stern, University of Colorado, Boulder; Ravindra Sudan, Cornell University; Roscoe White, Princeton Plasma Physics Laboratory; Scott Wilks, Lawrence Livermore National Laboratory; David Wineland, National Institute of Standards and Technology, Boulder, Colorado; Masaaki Yamada, Princeton Plasma Physics Laboratory; Michael C. Zarnstorff, Princeton Plasma Physics Laboratory.

Contents

PART II

❖

TOPICAL AREAS

CONCLUSION

APPENDICES

Plasma Science

From Fundamental Research to Technological Applications

Executive Summary

Plasma science is the study of the ionized states of matter. Most of the observable matter in the universe is in the plasma state. Plasma science includes plasma physics but aims to describe a much wider class of phenomena in which, for example, atomic and molecular excitation and ionization processes and chemical reactions can play significant roles. The intellectual challenge in plasma science is to develop principles for understanding the complex macroscopic behavior of plasmas, given the known principles that govern their microscopic behavior.

Plasmas of interest range over tens of orders of magnitude in density and temperature—from the tenuous plasmas of interstellar space to the ultradense plasmas created in inertial confinement fusion, and from the cool, chemical plasmas used in the plasma processing of semiconductors to the thermonuclear plasmas created in magnetic confinement fusion devices. A healthy plasma science enterprise can be expected to make many important contributions to our society for the foreseeable future. The purpose of this report is to provide guidance regarding the ways in which plasma science can contribute to society and to recommend actions that will optimize these contributions.

FINDINGS

1. Plasma science impacts daily life in many significant ways. It plays an important role in plasma processing, the sterilization of medical products, lighting, and lasers. Plasma science is central to the development of fusion as an

energy source, high-power radiation sources, intense particle beams, and many aspects of space science.

2. Plasma science is a fundamental scientific discipline, similar, for example, to condensed-matter physics. This fact is apparent when one considers the commonality of the intellectual problems in plasma science that span the wide range of applications to science and technology. Despite its fundamental character, plasma science is frequently viewed in the academic community as an interdisciplinary enterprise focused on a large collection of applications. Experiment, theory, and computation are all critical components of modern plasma science.

3. While the applications of plasma science have been supported by the federal government, no agency has assumed responsibility for basic research in plasma science. In general, there is a lack of coordination of plasma science research among the federal agencies.

4. As the development of plasma applications has progressed, small-scale research efforts have declined, particularly in the area of basic plasma experiments. This decline has led to a significant backlog of important scientific opportunities. This core activity in fundamental plasma science, carried out by small groups and funded by principal-investigator grants, is dangerously small, considering its importance to the national effort in fusion energy and other applied programs.

5. Plasma scientists in academic institutions are less likely to be in tenure-track positions than are other physicists, and courses in plasma science are currently unavailable at many educational institutions.

CONCLUSIONS

1. Plasma science can have a significant impact on many disciplines and technologies, including those directly linked to industrial growth. To properly pursue the potential offered by plasma science, the United States must create and maintain a coherent and coordinated program of research and technological development in plasma science.

2. Recognition as a distinct discipline in educational and research institutions will be crucial to the healthy development of plasma science.

3. There is no effective structure in place to develop the basic science that underlies the many applications of plasmas, and if the present trend continues, plasma science education and basic plasma science research are likely to decrease both in quality and quantity. If nothing is done by the federal government, it is likely that research in basic plasma science will cease to exist, and progress in the applications that depend on it will eventually halt.

4. The future health of plasma science, and hence its ability to contribute to the nation's technological development, hinges on the revitalization of basic plasma science and, in particular, on the revitalization of small-scale basic plas-

ma experiments. With regard to theory and modeling, although the current programs have been successful, there is a need for individual-investigator-led research on questions fundamental to basic plasma science.

5. Coordination of research efforts is vital, to make the most effective use of resources by maintaining complementary programs and to ensure that all critical problems are addressed.

6. Because of the commonality underlying all areas of plasma science, renewed emphasis on basic plasma science will benefit all areas. Therefore, it is appropriate that redistribution of funding to support basic plasma science come from all areas of plasma science.

RECOMMENDATIONS

1. To reinvigorate basic plasma science in the most efficient and cost-effective way, emphasis should be placed on university-scale research programs.

2. To ensure the continued availability of the basic knowledge that is needed for the development of applications, the National Science Foundation should provide increased support for basic plasma science.

3. To aid the development of fusion and other energy-related programs now supported by the Department of Energy, the Office of Basic Energy Sciences, with the cooperation of the Office of Fusion Energy, should provide increased support for basic experimental plasma science. Such emphasis would leverage the DOE's present investment in plasma science and would strengthen investigations in other energy-related areas of plasma science and technology.

4. Approximately $15 million per year for university-scale experiments should be provided, and continued in future years, to effectively redress the current lack of support for fundamental plasma science, which is a central concern of this report. Furthermore, individual-investigator and small-group research, including theory and modeling as well as experiments, needs special help, and small amounts of funding could be life-saving. Funding for these activities should come from existing programs that depend on plasma science. A reassessment of the relative allocation of funds between larger, focused research programs and individual-investigator and small-group activities should be undertaken.

5. The agencies supporting plasma science should cooperate to coordinate plasma science policy and funding.

6. Members of the plasma community in industry and academe should work aggressively for tenure-track recognition of plasma science as an academic discipline, and work with university faculty and administrators to provide courses in basic plasma science at the senior undergraduate level.

Additional recommendations regarding specific areas of plasma science are made in the main text of the report.

PART I

❖

Overview

INTRODUCTION

Plasma science is the study of the ionized states of matter. Plasmas occur quite naturally whenever ordinary matter is heated to a temperature greater than about 10,000°C. The resulting plasmas are electrically charged gases or fluids. They are profoundly influenced by the long-range Coulomb interactions of the ions and electrons and by the presence of magnetic fields, either applied externally or generated by current flows within the plasma. The dynamics of such systems are complex, and understanding them requires new concepts and techniques.

Plasma science includes plasma physics but aims to describe a much wider class of ionized matter in which, for example, atomic, molecular, radiation-transport, excitation, and ionization processes, as well as chemical reactions, can play significant roles. Important physical situations include partially ionized media and the interaction of plasmas with material walls. Thus plasma science draws on knowledge and techniques from many areas of science, including chemistry, fluid dynamics, and large-scale numerical computation, to achieve an accurate description of plasma behavior.

The goal of *plasma physics* is to describe elementary processes in completely ionized matter. In common with such fields as chemistry, condensed matter physics, and molecular biology, plasma physics is founded on well-known principles at the microscopic level. Description of plasmas typically involves use of Maxwell's equations for the electromagnetic fields and the Liouville or Boltzmann equations to model the dynamics of the electrons and ions, which are treated as point charges. Simpler approximations based on fluid descriptions for the electrons and ions (e.g., magnetohydrodynamics) are also used. The plasma medium is inherently nonlinear because the charged particles composing the plasma interact collectively with the electromagnetic fields produced self-consistently by the charge density and currents associated with the plasma particles.

Much of the basis for analyzing and treating plasmas has now been laid out, and a number of important advances in our understanding have been made. However, we are far from being able to make quantitative predictions of plasma behavior in many, if not most, of its manifestations. The intellectual challenge in plasma physics is to develop principles for understanding the complex macroscopic behavior of plasmas, given the known principles that govern their microscopic behavior.

The development of plasma science in the past three decades has been propelled by applications such as fusion energy, space science, and the need for a strong national defense, and this support has resulted in significant progress. Yet, by necessity, only those aspects that appeared to be more or less directly pertinent to applications received the lion's share of attention. Plasma science has benefited greatly from this support, but the field has now reached a level of maturity where many basic issues have been identified and remain to be resolved. Further progress will depend eventually on addressing these basic issues, rather than focusing only on the demands made on plasma science by applications. In turn, a greater understanding of the fundamentals of plasma science can be expected to advance significantly its successful application to the needs of society. Progress will be greatly inhibited without a strong experimental and theoretical research program directed at the fundamental principles of plasma science and not constrained to focus only on near-term applications. In particular, although theoretical and computational studies have spearheaded many of the advances in plasma physics in the past, well-planned and precisely controlled experiments will be crucial to further progress.

The panel was charged with the assessment of specific areas of plasma science that it refers to as *topical areas*. These include low-temperature plasmas, nonneutral plasmas, inertial and magnetic confinement fusion, beams, accelerators, and coherent radiation sources, and space and astrophysical plasmas. These areas vary in size, the nature of the scientific efforts, and the key scientific and organizational challenges facing them. Part II contains assessments of these topical areas with conclusions and recommendations specific to each. The panel was also charged with the assessment of *broad areas of plasma science:* basic plasma experiment, theory and computational plasma physics, and plasma science education; this is done in Part III. Although research and development in the topical areas is proceeding reasonably well, the panel's conclusion is that maintaining the vitality of basic plasma science faces severe difficulties unless there is concerted action by both the funding agencies and the scientific community. Because of the importance of present and potential applications of plasma science to our society, much benefit would be gained by a coherent program of support for basic plasma science. Much of the remainder of this overview chapter is devoted specifically to this issue, and the chapter concludes with a summary of the central messages of the report and the panel's general conclusions and recommendations.

THE ROLE OF PLASMA SCIENCE IN OUR SOCIETY

Plasma science impacts daily life in many significant ways. Low-temperature plasmas, in which electric fields in the plasma can impart significant energy to the electrons and ions but the plasmas are still cool enough to support a multitude of chemical reactions, are critical to the processing of many modern materials. This method of "plasma processing" is an enabling technology in the fabrication of semiconductors. Important applications include the plasma etching of semiconductors and the surface modification and growth of new materials. A recent National Research Council report,[1] which highlights the impor-

[1]National Research Council, *Plasma Processing of Materials: Scientific Opportunities and Technological Challenges*, National Academy Press, Washington, D.C., 1991.

tance of plasma processing in the electronics industry, indicates that the worldwide sale of plasma reactors alone amounted to $1 billion dollars in 1990 and is expected to double in the next five years. Other important uses of low-temperature plasmas include the "cold" pasteurization of foods, the sterilization of medical products, environmental cleanup, gas discharges for lighting and lasers, isotope separation, switching and welding technology, and plasma-based space propulsion systems.

Coherent radiation sources and particle accelerators rely on plasma concepts. Uses of intense electron beams include the bulk sterilization of medical products and food, toxic waste destruction via oxidation, processing of advanced materials, and new welding techniques. Free-electron-laser radiation sources have a variety of potential applications in medicine and industry, and they offer the possibility of intense, tunable sources of electromagnetic radiation in virtually all parts of the electromagnetic spectrum. Nonneutral plasmas in electromagnetic traps have application as ultraprecise atomic clocks and as a method to confine and manipulate antimatter such as positrons and antiprotons.

Plasma science is central to the development of fusion as a clean, renewable energy source. In order to control the fusion process, which is the source of energy of the Sun and the stars, we must learn to create hot, dense plasmas of deuterium and tritium in the laboratory. Great progress has been made toward this goal. Fusion-plasma confinement times have increased by a factor of more than 100 in the last two decades, and achievable temperatures have increased by a factor of 10. There is now in place an international collaboration to design the first prototype fusion power reactor, the International Thermonuclear Experimental Reactor (ITER). However, the continued refinement of the fusion concept and the optimization of fusion as a power source will require improved understanding of methods of confining and heating plasmas, as well as the development of techniques to lessen the damage to material walls due to the close proximity of the fusion-temperature plasmas. The leverage on investment in this area is tremendous. All major industrial nations have experienced a steady increase in the use of electricity—it is the energy type of choice. Nuclear fission plants are aging, fossil fuels continue to be of concern due to the production of greenhouse gases, and fusion offers the potential of large-scale electricity generation with abundant fuel supply and attractive environmental features.

We live in the 1% or so of the universe in which matter is not ionized, so plasmas are not readily apparent in our daily lives. However, as illustrated in Figure S.1, plasmas occur in many contexts, spanning an incredible range of plasma densities and temperatures. The most common examples of plasmas that we can actually see are the gas discharges in neon lights and the discharges in bolts of lightning. Most of the observable matter in the universe is in the plasma state (i.e., in the form of positively charged ions and negatively charged electrons). Plasma science provides one of the cornerstones of our knowledge of the Sun, the stars, the interstellar medium, and galaxies. We cannot understand such

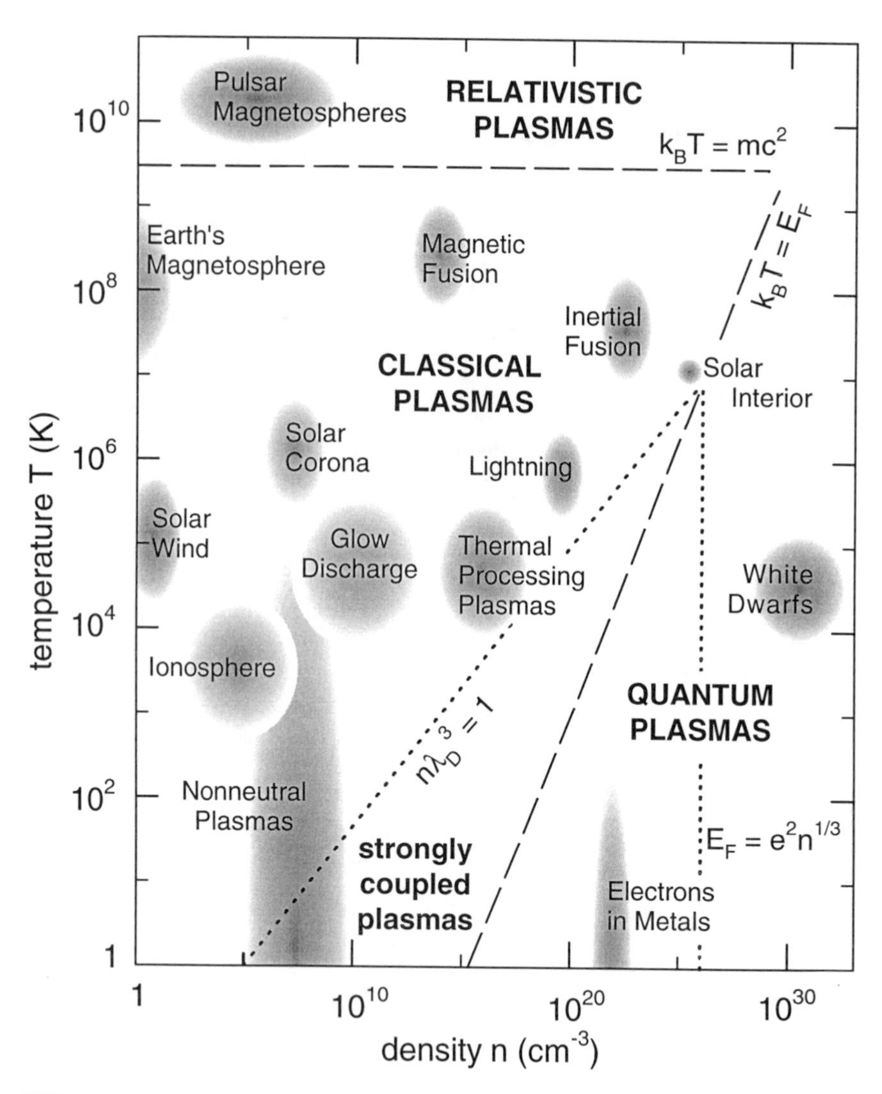

FIGURE S.1 Plasmas that occur naturally or can be created in the laboratory are shown as a function of density (in particles per cubic centimeter) and temperature (in kelvin). The boundaries are approximate and indicate typical ranges of plasma parameters. Distinct plasma regimes are indicated. For thermal energies greater than that of the rest mass of the electron ($k_BT > mc^2$), relativistic effects are important. At high densities, where the Fermi energy is greater than the thermal energy ($E_F > k_BT$), quantum effects are dominant. In strongly coupled plasmas (i.e., $n\lambda_D{}^3 < 1$, where λ_D is the Debye screening length), the effects of the Coulomb interaction dominate thermal effects; and when $E_f > e^2n^{1/3}$, quantum effects dominate those due to the Coulomb interaction, resulting in nearly ideal quantum plasmas. At temperatures less than about 10^5 K, recombination of electrons and ions can be significant, and the plasmas are often only partially ionized.

phenomena as sunspots, the formation of stars from interstellar gas clouds, the acceleration of cosmic rays, the formation and dynamics of energetic jets and winds from stars and galaxies, or the interaction of supernova remnants with interstellar gas, without the concepts of plasma science. Plasmas are central to many aspects of space science. The space plasma medium extends from the ionosphere surrounding the Earth to the far reaches of the solar system. "Space-weather" prediction in the ionosphere and magnetosphere is important for global communications, and the properties of space plasmas are important in determining the capabilities and longevity of spacecraft. Thus, although it is often not readily apparent, plasma science affects our society in a myriad of ways.

THE DISCIPLINE OF PLASMA SCIENCE

Common Research Themes

The panel has concluded that plasma science is frequently viewed, not as a distinct discipline, but as an interdisciplinary enterprise focused on a large collection of applications. The underlying, common, and critical feature of plasma science as a discipline is that its goal is to understand the behavior of ionized gases, and this requires fundamentally different techniques from those applicable to uncharged gases, fluids, and solids. This coherence of plasma science as a discipline is apparent when one considers some of the challenging intellectual problems, central to plasma science, that span applications in many of the topical areas. The impact of four such problems on the topical areas assessed in this report are summarized in Table S.1.

Wave-Particle Interactions and Plasma Heating

Understanding the interaction of plasma particles with the collective plasma oscillations and waves is a fundamental question with many practical applications. Basic scientific issues involve the trapping of particles in waves, the nonlinear saturation of wave damping, chaotic behavior induced in particle orbits, and particle acceleration mechanisms. Wave heating is an important method of heating fusion plasmas to the required temperatures for fusion. Waves can be used to drive electrical currents in plasmas. One promising scheme for a steady-state tokamak fusion reactor is to use waves to drive electrical currents to confine the plasma, instead of the present method of driving pulsed currents inductively. Wave-particle interactions are central to the operation of free-electron lasers and other coherent radiation sources, to many advanced accelerator concepts, and to methods of creating and heating low-temperature plasmas for plasma processing applications. Wave-particle processes are important in Earth's magnetosphere and ionosphere, and shock waves are the dominant production mechanism for cosmic rays of astrophysical origin.

TABLE S.1 Applications of Basic Plasma Research, Illustrating the Commonality of Scientific Issues Across Topical Areas

	Scientific Issue			
Topical Area	Wave-Particle Interactions and Plasma Heating	Chaos, Turbulence, and Transport	Sheaths and Boundary Layers	Magnetic Reconnection and Dynamos
Low-temperature plasmas	Magnetrons Plasma sprays	Instabilities in plasma processing	Plasma processing Lighting	Plasma torches MHD drag reduction
Nonneutral plasmas	ICR mass spectrometry	Precision clocks Fluid flows Antimatter storage	Switches Diodes	—
Inertial confinement fusion	Parametric instabilities Preheating	Turbulent mixing Rayleigh-Taylor instability	Plasma-driver interface	ICF plasma magnetic fields
Magnetic confinement fusion	rf current drive rf plasma heating	Energy and particle loss	Divertor operation Plasma-limiter interaction	Sawteeth and current profile dynamics in tokamaks Reversed-field pinches

Beams, accelerators, and coherent radiation sources	FELs Advanced accelerators	Instabilities in FELs and advanced accelerators	Cathodes	—
Space plasmas	Magnetosphere Ionosphere	Cometary and planetary atmospheres Solar wind	Magnetospheric and ionospheric boundaries Plasma-satellite interaction	Solar interior and corona Magnetopause
Astrophysical plasmas	Cosmic-ray acceleration	Accretion disks Dynamo viscosity	Pulsar magnetospheres Accretion disk boundaries	Solar and stellar magnetic fields

NOTE: ICR = ion cyclotron resonance
rf = radio-frequency
FEL = free-electron laser
MHD = magnetohydrodynamic
ICF = inertial confinement fusion

Chaos, Turbulence, and Transport

Most plasmas of interest are nonuniform in density and temperature, which results in the excitation of turbulent waves and fluctuations. These fluctuations in turn produce the transport of particles and energy that tend to drive the plasma toward a more uniform state. Very generally, turbulence and turbulent transport are not understood. The recent renaissance in nonlinear dynamics and studies of phenomena such as chaos provide new tools with which to attack these problems, and in fact, plasmas offer a convenient and often unique medium in which to study turbulent phenomena. Many of the scientific issues are now clear, and the plasma applications are many. For example, turbulent transport is the dominant mechanism for energy and particle transport in tokamak fusion plasmas. Turbulent transport is frequently the dominant particle and energy loss mechanism in low-temperature plasmas. It is important in the Earth's magnetosphere, in stellar convection zones, and in astrophysics in settings such as the interstellar medium.

Plasma Sheaths and Boundary Layers

Understanding the boundaries of plasmas (called sheaths) is a well-defined problem with many practical consequences. In magnetically confined fusion plasmas, the hot plasma cannot be allowed to contact the material walls. The result is that there must be large gradients in plasma temperature and density in the plasma and, frequently, non-equilibrium particle distributions. The precise character of these boundary layers can greatly influence the character of the bulk plasma and the rate at which wall damage occurs. Plasma sheaths are important in the plasma processing of materials. This sheath is adjacent to the material surface to be processed; therefore, the properties of this layer determine the characteristics of the plasma-matter interaction. A similar phenomenon occurs in space plasmas, where a plasma sheath separates a satellite from the surrounding space plasma, and the properties of this sheath determine the interaction of the plasma particles with the satellite. This is relevant for considerations such as surface damage and electrical phenomena. A type of boundary layer called a "double layer" can separate regions of plasma with distinctly different properties. Double layers are known to occur both in laboratory and in space plasmas, where they play an important role in determining the global configuration of the plasma. Related electrode-sheath phenomena are the least well understood aspects of lighting plasmas, and learning to control them better would lead to more efficient and longer-life products.

Magnetic Reconnection and Dynamo Action

The behavior of a magnetized plasma is determined largely by the configuration of the magnetic field in the plasma. Currents and plasma flows can induce changes in the topology of the field by breaking and reconnecting the magnetic field lines. This process occurs in (magnetohydrodynamic) "sawtooth" oscillations in tokamak plasmas, in sun spots, and in many astrophysical plasmas.

Dynamo action is the process by which a flowing plasma converts mechanical energy into magnetic field energy. This process is thought to be the origin of Earth's magnetic field, and it is likely to be one of the mechanisms for producing astrophysical magnetic fields. Very little is understood about magnetic reconnection and dynamo action, yet new techniques are now available to address these problems, by analytic methods, by computer simulation, and by suitably designed laboratory experiments.

These and other forefront problems in plasma science are described in Part III, and the relationship of this research to specific topical areas and applications is discussed in Part II.

Research and Education in Plasma Science

The findings and conclusions regarding the three broad areas of plasma science assessed in Part III are discussed in this section.

Basic Plasma Experiments

Of any of the topics in the panel's study, basic plasma experiments constitute the area of greatest concern. Progress in the physical sciences has relied historically on the close interplay between theory and experiment. Perhaps nowhere is this more true than in many-body physics, which naturally includes plasma physics. Physical phenomena can be identified, isolated, and studied most efficiently, quickly, and economically in experiments specifically tailored for this purpose. There are many advantages of basic experiments, compared to experiments done in settings determined by other considerations such as particular applications. These advantages include the flexibility to choose the setting to isolate a particular physical phenomenon, the ability to explore the broadest range of plasma parameters, and the ability to make experimental changes quickly, guided by the internal logic of the underlying science and by new results as they unfold.

Despite the importance of basic plasma experiments to plasma science, there have been clear warning signs for more than a decade of a deficiency in this area. This was expressed clearly in the Brinkman report, *Physics Through the 1990s*, written almost a decade ago.[2] The finding of the panel is that this situation has worsened since the Brinkman report was issued. There are several causes of this

---

[2]National Research Council, *Plasmas and Fluids*, in the series *Physics Through the 1990s*, National Academy Press, Washington, D.C., 1986.

problem, which include the narrowing focus of the large applied programs such as fusion and space applications. The future health of plasma science as a discipline hinges on the revitalization of basic plasma science, particularly the revitalization of small-scale basic plasma experiments—the area of most rapid decline in the last 20 years. Plasma science is suffering from application without replenishment: With major emphasis on applying what is known and without maintaining the basic scientific effort, the "seed corn" is quickly disappearing.

Although support for basic plasma science has declined over the past two decades, there has been important progress, which provides an idea of the potential contributions that basic research can make. Important achievements include a deeper understanding of the interaction of plasma waves with plasma particles. Many important nonlinear plasma processes have been isolated and understood, including some aspects of double layers, which, as discussed above, are the nonlinear interfaces between regions of plasma with distinctly different plasma properties; the effects of ponderomotive forces in causing the reorganization of plasmas; the filamentation of electromagnetic radiation in plasmas; and some aspects of the reconnection of magnetic field lines and magnetic reconfiguration of plasmas. Recently, the effects of chaotic particle orbits on plasma behavior have begun to be addressed by laboratory experiments. Each of these phenomena has many potential applications, so that progress on one topic is likely to have an impact on several different areas of plasma science. Progress has also been made in developing new techniques for experimental plasma studies, including some developed specifically for plasma applications and others that have resulted from advances in technologies such as electronics and computing.

As the development of applications has progressed, the decline of basic plasma experiments over the past 20 years has led to a significant backlog of important opportunities for basic experiments. We lack a basic understanding of many aspects of Alfvén waves, relevant to both space and fusion plasmas. We lack a basic understanding of magnetic reconfiguration processes and a host of other important nonlinear phenomena, including the wave-plasma interactions, plasma sheaths and boundary layers in magnetized plasmas, and dynamo action, discussed above. Progress in understanding each of these phenomena has been hindered by the lack of basic experiments designed to address key issues.

To reinvigorate experimental plasma science in the most efficient and cost-effective way, the panel concludes that emphasis should be placed on support for university-scale research programs. This conclusion is based on two findings: Many if not most of the important outstanding problems in basic plasma science can be addressed by this mode of experimentation. In addition, where small and individual principal-investigator-led programs are possible, the degree of flexibility, diversity, and creativity associated with this mode of research optimizes the limited resources that can be expected to be available for basic plasma research in the foreseeable future.

The panel estimates the size of the investment that will be required in the following way. Healthy, small-scale experimental research programs in plasma science typically involve a senior researcher, a postdoctoral researcher, and three to six graduate students. In addition, due to the nature of plasma experiments, an electronics technician is frequently required. The cost of such a group, including equipment purchases, typically ranges from $200,000 to $400,000 per year. (See Chapter 8 for details.) Given the erosion of the experimental infrastructure in basic plasma science, typical initial equipment purchases of from $300,000 to $600,000 are required for each new program. Based on the decline of basic experimental plasma science over the past 20 years, the panel estimates that a total of 30 to 40 new programs will be required to revitalize this area. Thus, the cost of such an effort would be approximately $15 million per year.

The panel's survey of the experimental research community indicates that the present number of efforts in the United States of the type described above, basic experimental plasma research, is less than 20. The required number of new groups (30 to 40) was estimated by considering the minimum-size scientific community that would provide sufficient cross-fertilization of ideas and stimulation, spread over this broad area of basic plasma science. Consideration of the size of a research community necessary to cover the existing range of forefront scientific problems (discussed in Part III) gives a similar estimate.

Theory and Computational Plasma Physics

Complementing experimental observation, theory, and computation are critical components of modern plasma science. The theoretical problems in plasma science are formidable. The goal is to achieve quantitative understanding of nonlinear, many-body phenomena in these nonequilibrium systems. Much progress has been made in this area in the past two decades, particularly in applied areas where there have been concentrations of effort, such as fusion, space plasmas, and plasmas relevant to defense applications. Examples include magnetohydrodynamic phenomena relevant to tokamak fusion plasmas, where the stability of these plasmas and the macroscopic dynamics of unstable plasmas are now quantitatively understood. Beyond such fluid models of plasma behavior, progress has been made in more detailed and powerful statistical mechanical descriptions of plasmas, including the effects of large particle orbits on plasma stability and the so-called gyrokinetic model of plasma dynamics. Many other nonlinear plasma problems have been addressed successfully, including the dynamics that results when an intense beam of electromagnetic radiation is incident upon a non-uniform plasma.

Fundamental understanding of a variety of coherent structures has been achieved—vortices and nonlinear phase-space correlations, for example. These concepts are of considerable significance because they have the potential to lead to simplified descriptions of otherwise complicated plasma behavior. Numerical simulations have provided deeper insights into a variety of plasma processes.

This progress has been driven by advances in computational techniques in general and also by the invention of new algorithms designed specifically for plasma computations.

Future challenges to theoretical and computational plasma physics arise from several different considerations. As mentioned above, the description of plasma behavior, given that plasmas are nonlinear and nonequilibrium many-body systems, typically presents the theorist with very difficult problems. In addition, a wide variety of parameter ranges and boundary conditions are relevant to plasmas of interest. Finally, the fact that there is a vast range of practical applications of plasmas—from plasma processing, to new radiation sources and accelerators, to fusion and global communications—requires not just a qualitative but a quantitative understanding of the underlying plasma behavior.

One particularly important and broad topic in theoretical and computational plasma science is the achievement of a deeper understanding of plasma turbulence and the associated transport that frequently results from it. Another broad area of fundamental importance is understanding the evolution of currents, flows, and magnetic fields in plasmas. These problems, which typically involve a breaking of the spatial symmetry, can lead to coherent structures and multiple length scales that are difficult to treat theoretically or numerically. Yet such phenomena are important in applications ranging from low-temperature plasmas for materials processing and welding to space and astrophysical plasmas.

While the current programs of theory and modeling have been successful, particularly in the context of the large applied programs, the panel concludes that two modes of research need to be reemphasized. There is a need for individual-investigator-led research on questions fundamental to basic plasma science, such as stochastic effects, novel analytical techniques, and a variety of nonlinear processes. Emphasis also needs to be put on pursuing the commonality of physical processes and mathematical techniques across the many subdisciplines of plasma science.

Education in Plasma Science

Plasma science is a fundamental scientific discipline that has made and continues to make significant contributions to our society. The field is intrinsically interdisciplinary, much like modern materials science. Yet, there needs to be a focus and home for plasma science in modern educational and research institutions for the field to develop properly and have the maximum impact. The panel has found that plasma scientists are less likely to be in tenure-track positions than are other physicists. Presently, courses in plasma science are unavailable at many educational institutions. If this trend continues, plasma science education and basic plasma science research are likely to decrease in both quality and quantity.

The entire field of plasma science, including its educational function, will be strengthened by the panel's recommendation that increased support of basic plasma science and university-scale research be provided by the National Science Foundation and the Department of Energy. The programs supported in this way can be expected to attract and maintain the scientific talent necessary to provide high-quality educational programs in plasma science and technology.

To develop adequate educational programs, given the breadth and diversity of plasma science, the panel makes the following recommendations to the plasma community. Since plasma science can contribute to a wide variety of scientific disciplines, educational programs should include courses tailored to the needs of areas such as low-temperature plasma physics and plasma chemistry, plasma processing, and astrophysical plasma physics, in addition to the major programs of fusion and space plasma physics. The panel believes that these programs would be helped by textbooks with the particular needs of these areas and relevant applications in mind. The panel suggests that senior-level courses in plasma physics become standard offerings in undergraduate curricula and that chapters on plasma physics be developed for more general science textbooks, to increase the level of plasma literacy of scientists and engineers outside the field.

SUMMARY OF TOPICAL AREAS

As described in the preface, the panel was charged with assessing specific topical areas of plasma science: low-temperature plasmas; nonneutral plasmas; inertial and magnetic confinement fusion; beams, accelerators, and coherent radiation sources; and space and astrophysical plasmas. This section contains an overview of these assessments.

Low-Temperature Plasmas

Low-temperature plasmas include those with many important technological applications, such as the plasma processing of materials for electronics, "cold" pasteurization of foods and sterilization of medical products, environmental cleanup, gas discharges for lighting and lasers, isotope separation, switching and welding technology, and plasma-based space propulsion systems. As an example of the impact of these technologies on industry and business, the use of plasma processing of semiconductors for electronics was recently reviewed by a separate National Research Council study.[3] The annual sales in equipment for the plasma processing of semiconductors amounted to $1 billion per year in 1990 and are projected to grow to $2 billion in 1995.

Low-temperature plasma science requires an understanding of atomic and molecular physics, plasma chemistry, and plasma physics. Many fundamental

---

[3]National Research Council, *Plasma Processing of Materials: Scientific Opportunities and Technological Challenges*, National Academy Press, Washington, D.C., 1991.

low-temperature plasma phenomena are poorly understood. Yet modern techniques are available to address a range of problems of fundamental importance that have important practical applications. Examples include the physics and chemistry of plasmas at material boundaries (i.e., plasma "sheaths"), the creation of plasmas by electrodeless discharges, and the stability and reproducibility of plasma discharges.

There are many important applications of low-temperature plasma science, yet there is no structure in place to support the fundamental research in this area that will be required to systematically develop these applications. Federal agencies have traditionally had only modest efforts in low-temperature plasma research; and recently, they have deemphasized these programs. Industry has usually engaged only in projects for which there is a short-term payoff. In contrast, there is an active effort in this area in Japan, and a large effort in low-temperature plasma research has recently been created in France. The crucial problem regarding low-temperature plasma science in the United States is the lack of a coordinated governmental program in this technologically important area. The panel recommends the creation of a coordinated support structure for fundamental research in low-temperature plasma science.

Nonneutral Plasmas

Nonneutral plasmas include pure electron plasmas and pure ion plasmas in electromagnetic and electrostatic traps, electron beams, and ion beams. Examples of applications of nonneutral plasmas are electron beams and plasmas for the generation of electromagnetic waves, pure ion plasmas in traps for atomic clock applications, advanced concepts for particle accelerators, and the confinement of antimatter such as positrons and antiprotons.

Nonneutral plasmas are more easily confined than neutral plasmas. Consequently, they can be more easily controlled and studied. Important questions that have recently been addressed include issues of plasma confinement, the creation of thermodynamic equilibrium states and controlled departures from equilibrium, and the mechanisms for the transport of particles and energy. Many of the concepts developed in the study of nonneutral plasmas have wider applications to understanding the physics of neutral plasmas and to fluid dynamics and atomic physics.

Nonneutral plasma physics is one area of basic plasma science that has progressed dramatically in the past two decades, and questions of fundamental importance have been addressed that are also relevant to technological applications. This was due, in large part, to a program of dedicated support for research in this area by the Office of Naval Research. This successful support of experimental and theoretical research on nonneutral plasmas should be used as a model for a program of renewed support of basic experiments in neutral plasmas that is recommended elsewhere in this report.

Nonneutral plasma research will be an important area for the foreseeable future, both from the point of view of fundamental plasma science in neutral and nonneutral plasmas and in exploiting this knowledge for technological applications. The panel recommends that continued support be given to research in this area and to the development of technological applications.

Inertial Confinement Fusion

The goal of inertial confinement fusion is to harness fusion power using intense lasers or ion beams to compress fusionable material, such as deuterium and tritium. The required plasma parameters for inertial confinement fusion are densities as much as 100 times that of ordinary matter and temperatures in excess of 100 million Kelvin. Much progress has been made in this area in the last decade. Important, new diagnostic techniques have been developed. The largest laser project, Nova, has conducted experiments to study the important problem of the physics of interpenetrating materials at accelerated interfaces. Computer simulations have clarified the role that fluid dynamical instabilities play in the dynamics of target compression. However, many challenging problems remain to be addressed. Examples include understanding stimulated Raman and Brillouin instabilities in laser-plasma interactions, particularly at high laser intensities, and understanding nonlinear plasma instabilities and the equations of state and opacity of matter at high densities and temperatures.

Many of the outstanding problems in this area have a high degree of commonality with important problems in other areas of science. Examples relevant to space physics and fusion include questions of plasma turbulence, particle acceleration and heating by electromagnetic radiation, and the effects of spatial inhomogeneities on wave propagation and mode conversion. Other important problems have much in common with optical science. Research in inertial confinement fusion can also benefit other fields. Examples include the development of short-pulse and x-ray lasers.

While progress toward inertial confinement fusion has been good, continued emphasis on programmatic milestones could leave unaddressed fundamental scientific questions crucial to the achievement of future goals. It is important for the program to reemphasize a broad-based program of support for relevant areas of basic research.

The commonality of scientific problems with other areas of science could be used to facilitate progress, both in the inertial fusion program and in related areas. Much of the fusion target program in the United States had been classified for security reasons. Much benefit should be gained by declassification currently being done by the Department of Energy. Given the commonality of problems in this area with those in other areas of plasma science and the importance of basic research in related fields to the achievement of fusion, the panel recommends that some resources be reallocated within the program to support the study of basic science relevant to inertial confinement fusion. Further, the panel recommends that the use of inertial confinement fusion facilities by scientists, working outside the program but on relevant problems, be encouraged.

Magnetic Confinement Fusion

Magnetic confinement fusion continues to be the largest driver for the intellectual development of plasma science. Central to the achievement of fusion in magnetically confined plasmas is the ability to confine hot plasmas (i.e., those with temperatures of more than 100 million Kelvin). Since these plasmas must eventually come in contact with material boundaries, this program also involves important considerations concerning low-temperature plasma science.

Much progress has been made in this field over the past two decades. Confinement times of fusion plasmas have increased by a factor of more than 100, and achievable temperatures have increased by a factor of 10. Progress has been made in the development of new diagnostics of plasma behavior, and these diagnostics have, in turn, led to a deeper understanding of the behavior of fusion plasmas. New methods have been developed to heat fusion plasmas and to drive electrical currents in these plasmas noninductively using intense neutral beams and radio-frequency electromagnetic waves. These methods of current drive could eventually permit the operation of a steady-state fusion reactor. New operating regimes with improved plasma confinement have been discovered, such as the so-called "high-confinement" and "very-high-confinement" modes. There has been progress in the understanding of plasma stability as well as in understanding the interface between the plasma edge and the material walls of the confinement vessel.

A key element in the magnetic confinement fusion program is the development of the International Thermonuclear Experimental Reactor (ITER). This device will be designed to test elements of reactor-relevant plasma science not possible by other means such as the physics of ignition. However, there are many other plasma processes relevant to controlled fusion that will not be able to be addressed effectively by the ITER program. The physics of the edge plasmas in tokamaks needs to be better understood. Advanced modes of tokamak operation at very long pulse lengths will be studied in the Tokamak Physics Experiment (TPX), now planned as a national facility for such studies. Finding improved methods of removing large quantities of heat from the plasma edge is an immediate problem. The efficient production of self-generated plasma currents by high plasma pressures (so-called "bootstrap currents") is an important goal of advanced tokamak configurations that is not likely to be studied efficiently in the ITER program. Experiment and theory should continue in the search for optimized geometries and operating conditions to improve reactor efficiency and power-handling capabilities.

Crucial to the operation of a fusion device is the transport of particles and energy by plasma turbulence, and turbulent transport has been the dominant transport mechanism in all magnetically confined fusion plasmas to date. There is, as yet, only an extremely limited first-principles understanding of the turbulence in fusion plasmas and the resulting transport. Any predictive capability that does exist is based on empirical "scaling laws" that must be validated when applied outside the operating parameter range of present and past fusion devices. A quantitative understanding of this transport and the ability to control it could potentially lead to improved reactor performance and reduced size and cost.

This fundamental base of plasma science is crucial not only for the efficient development of a successful fusion reactor, but also for quantitative understanding of fusion-related plasma science, which will continue to be important in maximizing the competitiveness of fusion power in the decades to follow.

The panel recommends that there be established a coordinated research program in fusion-relevant plasma physics. This will require a range of project sizes, in order to optimize the particular experiments to study the relevant plasma processes. Experimental research is most efficiently done on the smallest scale possible. This allows the greatest flexibility in making changes, as required by new results and discoveries, as well as the greatest exploration of the relevant parameter ranges at minimum cost. Many fundamental questions in basic plasma science should be addressed by small experiments that, in many cases, are specifically designed for a particular purpose. Other questions can be addressed only in larger devices. To study the effects of fusion products (e.g., alpha particles at energies of a few million electron volts) on fusion plasmas, reactor-sized devices, such as the Tokamak Fusion Test Reactor or the Joint European Torus, are required. Thus, a coordinated program of fusion plasma research will require a range of devices and programs, from small, basic experiments that isolate and address fundamental questions in plasma science to experiments on the largest fusion devices.

If the present trend toward large experiments continues without adequate attention paid to a broader base of experimental research facilities, a dangerous gap will develop in our ability to address the wide range of questions important to fusion-relevant plasma physics. Many important questions in fusion plasma physics might be more appropriately addressed by smaller, long-term research programs dedicated to isolating and studying fundamental plasma phenomena in a more complete and systematic manner.

The panel recommends that the program in magnetic confinement fusion include support for a range of projects, with the sizes chosen to best suit the particular plasma problem. Provision needs to be made for research on fusion-relevant basic plasma science. The details of this recommendation are given below in Part II.

Beams, Accelerators, and Coherent Radiation Sources

Until now, progress in this area has been driven by many important defense applications. Given recent changes in world politics, the need for such programs has greatly decreased. Military applications aside, this area of plasma science has a wide variety of important technological applications. Examples of uses of intense electron beams include the bulk sterilization of medical products and food, toxic waste destruction via oxidation, the processing of advanced materials, and new welding techniques. Free-electron laser radiation sources offer the possibility of providing intense, tunable sources of electromagnetic radiation in virtually all parts of the electromagnetic spectrum from the far infrared to x-ray wavelengths. They have a variety of potential applications in medicine and industry. The development of x-ray lasers is in its infancy but holds promise for many important practical applications.

To effectively pursue such applications will require a coordinated research and development effort. The panel recommends that beams, accelerators, and coherent radiation sources be given high priority for "defense conversion" funding.

Space Plasmas

Space plasma physics is concerned with the observation and understanding of naturally occurring solar-system plasmas. It is an evolutionary field, and progress has been achieved incrementally. The space plasma medium extends from the ionosphere of Earth to the far reaches of the solar system and encompasses plasmas of many types. Portions of this domain, such as the magnetosphere of Uranus, have experienced only brief, exploratory coverage, while others, like Earth's ionosphere and magnetosphere, have been investigated relatively thoroughly. In the case of the former, we are still at the stage of trying to deduce gross plasma structure from limited data; with the latter, we are in the process of understanding specific mechanisms that are responsible for the observed morphology. Occasionally, an entirely new physical situation is encountered, unlike anything previously observed in either space or in the laboratory, and this opens new scientific vistas. One example is the dusty plasmas of comets and planetary rings that are dominated by the dynamics of charged macroparticles for which gravitational and electromagnetic effects are of comparable importance.

There are many applications of plasma physics to space science, ranging from the development of plasma thrusters for spacecraft propulsion to "space weather" prediction in the magnetosphere and ionosphere, which has important consequences for physical phenomena on Earth, such as global communications. To some extent, space plasma physics draws upon the vast body of knowledge accrued through the laboratory program for the analysis, interpretation, and modeling of phenomena. Frequently, however, the parameters and the nature of boundary conditions are such as to render the space plasma physics unique, and necessitate entirely new theory or computer modeling. Thus, space plasma physics also contributes to the expansion of our knowledge of basic plasma physics. For example, our extensive understanding of collisionless shocks is based largely on space plasma studies of the Earth's bow shock. In the past, space plasmas have also been used as media in which to study phenomena of both applied and intrinsic interest and importance. However, these aspects of space plasma physics have now been deemphasized programmatically to the point of virtual extinction.

Observation is central to space plasma physics. Although observations are expensive to make, especially those requiring spaceflight, advancement in the field will continue to rely heavily on carefully planned and judiciously selected experiments to provide data that underlie new and refine old ideas. Technological improvements in detection systems and data handling capabilities can be expected to provide increasingly complete and accurate data on which to base models and theories. Recent progress in this area has been impressive, and the prospects for the future are very good.

The ambient space plasma can be modified by a number of techniques, including the injection of waves and particle beams, the injection of plasma and neutral gas, and perturbation by space vehicles. Such perturbations provide opportunities to isolate and study space plasma effects in detail and to create space plasmas relevant to other regions of space. Of particular concern to the panel is the fact that programs in this area of active, space plasma experimentation have recently been deemphasized by the funding agencies, and the panel recommends that this trend be reversed.

Given the spatial and temporal intermittencies of space plasma measurements, a program in laboratory experiments to study space plasma phenomena could be of great benefit. Such experiments have been supported in the past only to an extremely limited degree, due in large part to the fact that the design of experiments with appropriate scaling to space conditions is difficult in laboratory-sized devices. Advances in laboratory plasma experimentation have now progressed to the point that relevant plasma processes can be investigated in the laboratory with a degree of control, precision, and repeatability not achievable in situ. The panel recommends that an initiative be created for the support of laboratory experiments relevant to space plasmas.

Understanding space plasma phenomena frequently requires a combination of extensive data analysis, theory, modeling, and laboratory experiments, in addition to in situ observation. There is concern that in response to the pressure of escalating costs for observations, support for these other aspects of space plasma science has shrunk to unhealthy levels. The panel recommends that NASA and NSF fund a vigorous observational program, including both in situ and ground-based facilities, properly balanced with complementary programs of theory, modeling, and laboratory experiments.

Astrophysical Plasmas

Plasma physics is relevant to almost every area of astrophysics, ranging from stellar and interstellar plasmas to star clusters and galaxies. The potential contribution of plasma physics to astrophysics has been limited by the fact that plasma physics is not yet part of the standard graduate astrophysics curriculum, and this should be changed. Also of concern to the panel is the fact that plasma astrophysics does not have a distinct home in any of the federal funding agencies.

Examples of plasma astrophysics where there has been significant progress include models of magnetized accretion disks and associated jets and winds, including the effects of relativity, strong magnetic fields, rapid rotation, and magnetohydrodynamic waves and instabilities. Mechanisms of particle acceleration in plasma shockwaves have been clarified that are relevant to the acceleration of cosmic rays in the interstellar medium. Models have also been developed to describe the convective fluid motion in stars, including crucial effects arising from the presence of strong magnetic fields on the flow of stellar material.

While there has been progress in plasma astrophysics, many fundamental problems need to be addressed. These include the description of dense stellar plasmas with temperatures in excess of 20 million Kelvin and densities 10 to 100 times solid density. Other important problems involve turbulent plasmas, the origin of the magnetic fields in the universe, and magnetic field line reconnection.

Plasma astrophysics is not yet recognized as a coherent discipline by any federal funding agency. Yet, plasma astrophysics deals with phenomena that are important to virtually all aspects of astronomy and astrophysics, and many of these problems are central to basic plasma physics. The panel recommends that there be established interdisciplinary programs in the National Aeronautics and Space Administration and in the National Science Foundation to fund research in plasma astrophysics, including research on basic plasma processes relevant to astrophysical systems but not tied to any particular application.

CENTRAL MESSAGES OF THIS REPORT

The panel was charged with assessment of the state of plasma science in the United States and evaluation of its potential to contribute to the technology base of our society. It was further charged with assessing the institutional infrastructure in which plasma science is conducted, identifying changes that would improve the research and educational effort, and making recommendations to federal agencies and to the community to address these issues.

The theme of this report is that, although plasmas are pervasive in nature and many of the applications of plasma science are being pursued and exploited effectively, plasma science is not adequately recognized as a discipline. Consequently, there is not an effective structure in place to develop the basic science that underlies its many applications. The potential contributions of plasma science to our society would benefit greatly by a coordinated effort of support for fundamental research, not tied to specific programs but designed to establish this basic scientific foundation.

This report describes many of the developments in theory and experiment that have led to important industrial applications, significant commercial and residential uses, and a deeper understanding of the universe. While delineating the many successes and identifying the exciting and potentially critical challenges, this report is an expression of concern. Applications depend on development, which in turn depends upon a fundamental understanding of the underlying science—the sine qua non of new development. It is the view of experts in the field that this fundamental understanding is, in most cases, best obtained by individual and small-group experimental and theoretical efforts, typical of university-scale programs. The problem is not that there is a gross imbalance in the total funding going into plasma science and technology, but that there has been a gradual, long-term decrease in support for fundamental research in plasma science. The result is that there is now a clear need for such support, particularly in the areas of small-scale, basic plasma experiments and complementary small-group theoretical efforts.

As discussed above, a wide variety of programs pursue the applications of plasma science, including those in space, fusion, and the plasma processing of materials; yet there is great commonality of the underlying science. Even in the context of a particular application, there are often several programs, frequently spread across more than one federal agency. More often than not, this diversity is justified and healthy. However, coordination of research efforts is vital to eliminate duplication, to make the most effective use of resources by maintaining complementary programs, and to ensure that all of the critical problems are being addressed.[4]

A prime example of the existing lack of coordination in plasma research is the fact that *no* agency or agencies have yet assumed the responsibility for basic research in plasma science.

Thus, the central messages of this report are threefold:

1. Small-scale research provides much of the fundamental base for plasma science;

2. Such individual-investigator and small-group research is in need of support; and

---

[4]One beneficial effort of this type is the recent, informal coordination of experimental space plasma research by the National Aeronautics and Space Administration, the Office of Naval Research, and the National Science Foundation.

3. There is a need for increased coordination of federally funded plasma science research.

These conclusions coincide with the principal findings and recommendations of the Brinkman report,[5] which was an overall assessment of the future of plasma physics in the United States:

> Direct support for basic laboratory plasma-physics research has practically vanished in the United States. The number of fundamental investigations of plasma behavior in research centers is small, and only a handful of universities receive support for basic research in plasma physics. A striking example is the minimal support for basic research in laboratory plasmas by the National Science Foundation. . . . Because fundamental understanding of plasma properties precedes the discovery of new applications, and because basic plasma research can be expected to lead to exciting new discoveries, increased support for basic research in plasma physics is strongly recommended.

If anything, the state of basic plasma science has worsened in the nine years since the Brinkman report was published. This situation can be remedied only by the creation of a coherent and coordinated plan for the support of basic plasma science.

CONCLUSIONS AND RECOMMENDATIONS

Ongoing research and development programs in the United States have produced important advances in plasma-related science and technology. Plasma science holds promise of further progress in the future, including advanced methods of processing materials, better methods for cleaning up environmental hazards and mitigating the effects of deleterious chemicals, new methods of accelerating particles and producing electromagnetic radiation, progress toward fusion energy, and improved understanding of our space environment and the astrophysical media of the universe. Thus, plasma science can have a significant impact on many disciplines and technologies, including those directly linked to industrial growth. This impact, however, is critically dependent on the support of basic plasma science. It will be important to effect some shift of research funds to this area because of its close relation to applications.

To properly pursue its potential, the United States must create and maintain a coherent and coordinated program of research and technological development in plasma science. Currently, support for basic plasma science is mostly for small programs, found in many agencies, and not coordinated among agencies. The Department of Energy has large programs in the development of fusion energy, by both magnetic and inertial confinement schemes, but it has no unit in

[5]See footnote 2, p. 15.

the Basic Energy Sciences Division to provide support for the study of the fundamentals of plasma science. The National Science Foundation (NSF) is viewed as the supporter of basic science in universities, and there are several quite small plasma programs scattered throughout the agency. However, basic plasma science has no identified home in NSF and, thus, no specified coordinating and review point and no sponsor.

Below, the panel recommends increased support of university-scale experimental research in basic plasma science in the amount of $15 million per year. The justification for this amount is discussed above and in Chapter 8, Basic Plasma Experiments. Although it may seem that this could have only a small influence on a field with an annual budget in excess of $400 million, the expenditure on other than the largest applications, fusion and space plasmas, is less than 10% of this amount.

Consequently, an investment of $15 million on basic experiments can be expected to provide an important stimulus to the entire field. It can also be expected to have a multiplicative effect in that the results in basic plasma research will provide the foundation for research more closely related to all of the applications, including space and fusion.

While many successful programs in plasma science are currently under way, there is a lack of support for the basic aspects of plasma research, particularly where the payoff to a specific program cannot be justified in the near term. The development of plasma science would be improved substantially by its recognition as a scientific discipline.

Given these findings and conclusions, the panel recommends the following six actions:

1. To reinvigorate basic plasma science in the most efficient and cost-effective way, emphasis should be placed on university-scale research programs.

2. To ensure the continued availability of the basic knowledge that is needed for the development of applications, the National Science Foundation should provide increased support for basic plasma science.

3. To aid the development of fusion and other energy-related programs now supported by the Department of Energy, the Office of Basic Energy Sciences, with the cooperation of the Office of Fusion Energy, should provide increased support for basic experimental plasma science. Such emphasis would leverage the DOE's present investment in plasma science and would strengthen investigations in other energy-related areas of plasma science and technology.

4. Approximately $15 million per year for university-scale experiments should be provided, and continued in future years, to effectively redress the current lack of support for fundamental plasma science, which is a central concern of this report. Furthermore, individual-investigator and small-group research, including theory and modeling as well as experiments, needs special help, and small amounts of funding could be life-saving. Funding for these

activities should come from existing programs that depend on plasma science. A reassessment of the relative allocation of funds between larger, focused research programs and individual-investigator and small-group activities should be undertaken.

5. The agencies supporting plasma science should cooperate to coordinate plasma science policy and funding.

6. Members of the plasma community in industry and academe should work aggressively for tenure-track recognition of plasma science as an academic discipline, and work with university faculty and administrators to provide courses in basic plasma science at the senior undergraduate level.

PART II

Topical Areas

1

❖

Low-Temperature Plasmas

INTRODUCTION

During the last half century, low-temperature plasmas have made a dramatic impact on society, significantly improved the quality of life, and provided challenging scientific problems. Examples are the fluorescent lights that can be found in almost every home in America; high-power switches that control the electrical grid of the United States and divert electrical power on command; gas discharge lasers, including the red He-Ne laser, which was the first gas laser invented, and the high-power, infrared, CO_2 lasers that are used daily in surgery and metal working; and plasma sources that provide positive and negative ions for ion-beam accelerators. These ion sources are used to implant ions into materials, including semiconductor chips for the computer industry, and to harden bearings to increase the life and reliability of high-performance engines. Provided the opportunity, the field of low-temperature plasmas will continue to make significant contributions.

Based on the preceding paragraph, it is not surprising that low-temperature plasmas are important in many disciplines. Typically, they are high-pressure collision-dominated plasmas that have average electron energies of 1-10 eV. The purity of the gas is often important, and the physics and chemistry of the excited atomic states dominate the discharge characteristics. In industrial applications, the stability of the discharge frequently impacts the design and utility of the process, and the heterogeneous wall chemistry often impacts its reproducibility and reliability.

Unfortunately, because basic research in this area has been neglected for many years, there is a severe lack of quantitative and experimental understanding of a wide range of phenomena that occur in low-temperature collision-dominated plasmas. Most low-temperature plasma applications involve complex reactions between electrons and a host of atomic, molecular, and ionic species. These species are found in highly excited states not encountered in nonplasma environments. Operation of plasmas in applications ranging from lasers to materials processing and lighting requires optimization of the densities of these species. Scientists modeling these systems require a broader range of diagnostics to characterize species densities in benchmark plasmas, and more powerful methods for measuring, calculating, or approximating the cross sections that dominate the rate equations.

In some cases, such as microwave breakdown, the positive column of dc metal-vapor rare-gas discharges, and wall-stabilized arcs, researchers have obtained experimental data, theoretical understanding, and predictive models. However, much of this basic research was performed before 1960. In some cases with immediate industrial and government applications the information was updated in the 1970s, using modern experimental and modeling techniques. Examples include fluorescent lamps, high-intensity lamps, electron-beam-controlled discharge lasers, some specific plasma processes, and arcs (e.g., in discharge-limiting situations, such as transport in weakly ionized swarms and near thermal equilibrium). This research produced a significant improvement in the performance of devices using these plasmas. Recent research was driven by interest in high-power lasers for ballistic missile defense. The decline of interest in that use has severely reduced related funding.

In other areas, there has been limited progress during the last 30 years, including understanding phenomena such as collisional discharges in magnetic fields in the presence of boundaries, transient discharges and sheaths, discharge stability, and plasma interactions with practical surfaces. For example, recently there has been much interest in the dc cathode fall, since modeling and experiments are much further ahead for bulk-phase plasmas than for cases, such as the cathode fall, in which plasma contact with surfaces is important.

Lack of research support in the physics of low-temperature plasmas has resulted in a low level of training in collision-dominated low-temperature plasmas and in the training of engineers and physicists for plasma processing. No federal agency claims responsibility for this area.

The existing support has emphasized short-term goals and work only on current government- and industry-related topics. In FY 1991, there were only two long-term projects, and neither is currently funded. It is our understanding that since the beginning of FY 1992, there has been essentially no low-temperature plasma research project with more than a one-year time scale, since research in this area is dominated by the current needs of the radio-frequency plasma processing and lighting industries. This short time scale severely discourages new, innovative, or thorough research. Novel experimental and modeling techniques should be developed to explore new areas and provide more quantitative work in existing areas.

A serious problem in low-temperature collision-dominated plasmas has been the lack of reproducible experimental verification of theoretical predictions. This is partly due to the critical dependence of the relevant phenomena on surface conditions and gas purity. It is also due to the fact that the models are too limited and qualitative to be tested or to be of general use. While gas purification techniques have been known for many years, the role of impurity effects in practical systems is often poorly understood. The problem of the reproducibility of practical surfaces is very difficult, and few useful and successful recipes exist. Even fewer techniques exist for characterizing practical surfaces with regard to their interactions with plasmas. As a result, most of the successful quantitative gas-discharge investigations are of phenomena that are relatively free of surface effects (i.e., microwave breakdown, the positive column, swarm transport, and near-equilibrium radiation). However, given adequate funds, more realistic models could be developed to investigate these complex phenomena with modern computer facilities.

The Japanese government has long supported an active program in basic gas discharge research, particularly in its engineering schools. Japanese research is recognized internationally for its quality and impact. In spite of a major focus on plasma processing, many Japanese faculty still devote a significant fraction of their time and resources to basic, undirected research. In recent years the French government also has supported a large effort in low-temperature collision-dominated plasma research, which has produced a large fraction of the invited papers at recent international meetings.

To change the situation in the United States will require strong support for research in applied physics and engineering in the area of the basic physics of low-temperature plasmas. Experimental programs emphasizing quantitative and reproducible results will be necessary to properly test the predictions of theoretical models. Improved understanding of these plasmas is necessary for applications such as plasma processing and environmental cleanup. This basic research can also be expected to yield innovative experimental techniques and novel modeling methods, and it will provide highly trained scientists and engineers in low-temperature plasma science.

That low-temperature plasmas are crucial in so many technologies is both a strength and a weakness. These plasmas are indispensable in today's highly technical world, but since they are useful in many apparently disconnected disciplines, no agency has taken responsibility for research in low-temperature plasmas.

This chapter focuses on the following important areas of low-temperature plasma physics: lighting, gas discharge lasers, plasma isotope separation, space propulsion, magnetohydrodynamics, and the use of plasmas for pollution control and reduction. Another major area is plasma processing, which was addressed in detail in the recent report of the National Research Council (NRC) Panel on Plasma Processing of Materials.[1] The findings and recommendations of that study are summarized below in the section "Plasma Processing of Materials."

LIGHTING

Lighting has been one of the principal areas contributing to the understanding of low-temperature plasmas, and historically it has been responsible for much of the low-temperature plasma research in industry. Unfortunately, due to the severe recession of the late 1980s and early 1990s, this research effort has declined rapidly. Westinghouse and GTE-Sylvania have sold their lighting divisions to foreign investors, leaving General Electric the only large U.S. lighting company. By contrast, low-temperature plasma research in the Far East has been increasing rapidly; it is now three to four times larger than that in the United States.

Important contributions from lighting in the past 10 years include the control and modification of the electron-energy distribution function and novel laser diagnostics that provide valuable microscopic information about discharge parameters. Other important contributions include sophisticated and predictive models of lighting discharges and an understanding of the effects of isotopic gas mixtures in low-pressure mercury rare-gas discharges.

Major technological innovations have been made in the last decade in many areas. They include lower-power compact fluorescent and high-intensity discharges (HID), a variety of electrodeless discharges such as microwave, rf, and surface wave discharges for practical lighting applications, and the electronic ballasting of light sources. Other important innovations include an improved understanding of heterogeneous chemistry, resulting in superior performance and better compatibility with existing and novel materials, and the development of a systems approach to light sources that integrates principles of plasma discharges, materials, electronics, and homogeneous and heterogeneous chemistry.

Although most of the R&D for lighting plasmas is performed by the lighting industry, the field has also benefited from advances in other disciplines. For example, solid-state, plasma processing, and materials advances made in other industries and in academic and research institutions have contributed to the progress in the lighting industry.

A fundamental understanding of many processes is necessary for the lighting industry to design and fabricate higher-efficiency lamps. Therefore, university and government research has and will continue to impact the industry. Areas of research include, for example, local thermodynamic equilibrium (LTE)

---

[1]National Research Council, *Plasma Processing of Materials: Scientific Opportunities and Technological Challenges*, National Academy Press, Washington, D.C., 1991.

and non-LTE low-temperature plasmas, sophisticated diagnostics, and measurement of atomic and molecular cross-section data for electron-impact processes of importance to the lighting industry.

Because industrial R&D goals have become increasingly short-term over the last 10 years, it is important that the academic and government communities initiate longer-term R&D in plasmas related to lighting. Examples of research and development that would be of great benefit to the lighting industry include the application of massive computation techniques to lighting problems and improved diagnostics that will provide detailed information about the behavior of lighting plasmas. New approaches to modeling and probing complex sheaths associated with thermionic electrodes, for example, would help the lighting industry reduce mercury and thorium usage and would also increase lamp efficiency and life. Other important topics include methods to obtain higher conversion efficiency of electrical energy into radiation and the evaluation and exploitation of solid-state sources for lighting applications.

Scientific opportunities in lighting plasmas include the exploration of novel ways of producing monoenergetic or narrow electron-energy distributions in discharges to selectively excite electronic states, resulting in the more efficient production of radiation and the reduction of long-wavelength emission and thereby enhancing visible emission, using the principles of quantum electrodynamics and quantum interference.

Radiation from lamps can have important applications in environmental cleanup and other areas, including water purification with light, promoting algae growth with special metal-halide sources to reduce heavy metal concentrations in water, accelerating food growth, and a variety of display applications.

Lighting plasmas are synergistic with the fields of plasma deposition and etching, materials science, electronics, and lasers. However, increased scientific productivity in this area will require new basic experimental facilities.

GAS DISCHARGE LASERS

The field of gas discharge lasers has had considerable government support over the last 40 years. Strong support in the 1970s and 1980s led to an improved understanding of the basic phenomena in high-pressure plasmas, including electron-impact excitation cross sections of vibrational and electronic excited states, the physics of the stability of high-pressure discharges, and the homogeneous chemistry and products of excited-state reactions. Advances in the understanding of discharge physics include improved understanding and predictive capability of the discharge parameters and the stability of the plasma, discovery of the dominant impact that excited states have on discharge physics and laser chemistry, and an increased knowledge of electronic kinetics and the interaction between secondary electrons and excited states.

This research made significant contributions to advancing the state of the technology by increasing the efficiencies of lasers from a fraction of a percent to 10%. This improvement was particularly dramatic in excimer lasers. Examples include the rare-gas lasers that radiate in the vacuum ultraviolet (VUV), the rare-gas halide lasers that lase in the ultraviolet (UV), the rare-gas triatomic excimers that have broadband emission in the visible, and the metal excimers that emit in the visible and UV.

Gas lasers have produced many important technological capabilities. Examples include optical lithography, where rare-gas fluoride lasers have extended the resolution to less than 0.5 μm; laser working of metals, where high power CO_2 lasers are now used routinely for welding, cutting, and marking in industry; and medicine, where lower-power gas lasers have made a significant impact, including the use of surgical CO_2 lasers and excimer lasers for treating eyes and occluded arteries.

The combined gas laser market is presently of the order of $300 million and is predicted to grow at an annual rate of approximately 5%. There are also other emerging uses for these lasers, such as LIDAR (laser radar) for airports that can measure the location of wind shear and thereby increase the safety of air travel, and laser-produced x-ray sources for microlithography.

PLASMA ISOTOPE SEPARATION

Funding for plasma isotope separation has decreased dramatically in the post-Cold War era. Isotope separation has been actively investigated for the last 20 years, principally by plasma centrifuge, laser (AVLIS), and ion cyclotron resonance techniques. Of these methods, the AVLIS program at Lawrence Livermore National Laboratory has been the most strongly supported. The results are classified. Briefly, the separation process involves the selective ionization of one isotope and the subsequent collection of this ion. Lasers are used to ionize the desired isotopes, which form a low-temperature plasma. Plasma physics issues that have to be addressed include excited and ionic species reactions, homogeneous chemistry, and the physics and chemistry of the sheath near the collection electrodes. More conventional plasma isotope separation has been investigated on a much smaller scale by several groups including the FOM Institute for Plasma Physics in the Netherlands, Yale University, the Max Planck Institute in Germany, the Sydney University School of Plasma Physics in Australia, the National Space Research Institute in Brazil, and TRW in the United States.

There are important uses of isotope separation besides nuclear fuels enrichment, including medical diagnostics, chemistry, and basic research. Thus, the development of plasma centrifuge technology offers a number of potential opportunities. The demand for stable, enriched isotopes for medical applications grows each year. The plasma centrifuge offers an improved means of meeting this need. However, at present, many important basic plasma phenomena remain to be understood in the rotating plasmas utilized in plasma centrifuges, and further research in this area will be necessary to fully exploit their potential. In addition to isotope separation, the plasmas developed for plasma centrifuges can also be expected to be useful for other applications, including use in imploding, "z-pinch," plasma x-ray sources and in plasma switches.

PLASMAS FOR ELECTRIC PROPULSION OF SPACE VEHICLES

Space electric propulsion has been studied for the last three decades. Concepts that have been investigated include expanding electrically heated plasmas, accelerating plasmas with thrusters and plasma guns, accelerating ions, and laser propulsion. Research groups have included TRW in Redondo Beach, California; Avco Everett Research Laboratory in Everett, Massachusetts; the National Aeronautics and Space Administration (NASA) Lewis Research Center; and several university research groups. Electric plasma propulsion requires less fuel mass than chemical systems, potentially making launching of satellites and space exploration less expensive. Decreasing the weight of fuel and hence the overall payload could have a significant impact on the $9.5 billion currently spent annually for launches: $5 billion by the Department of Defense (DOD), $3 billion by NASA, and $1.5 billion by industry.

A plasma propulsion device (an ion accelerator) has been tested and has worked successfully in space for 13 years. A key issue for any such device that is launched into space is its reliability and longevity in both its on-the-shelf and operating lives. This makes the use of electrodes problematic. TRW has been pursuing electrodeless thrusters, and these plasma accelerators potentially could satisfy the demanding reliability requirements of space qualifiable systems. Other areas that require research include a better understanding of the plasma, identification of the appropriate gas or fuel, and matching the electrical driver to the nonlinear plasma load. There is presently a pressing need for low-power (e.g., of order 100 W) thrusters for long-term maintenance of orbits. The requirements for interplanetary missions will require significantly higher-power thrusters. The development of space plasma propulsion systems also is synergistic with other applications. For example, plasma accelerators can be used in plasma processing and in the simulation of space plasmas to determine the chemistry of these reactive media on satellites.

MAGNETOHYDRODYNAMICS

The branch of magnetohydrodynamics (MHD) of interest here is that concerned with plasmas at low temperatures (2000-10,000 K) and high pressures (1-10 atm). Most of the current work in this area is engineering development rather than scientific research. These efforts have focused primarily on electrical power generation, with some effort directed toward space vehicle thrusters. The power generation community is mostly industrial, and much of this work has been performed by Textron Defense Systems (formerly the Avco Everett Research Laboratory). There is some university involvement, such as the work at the University of Tennessee Space Institute. Workers in the field are generally engineers with plasma and fluid, thermoscience, mechanical engineering, or electrical engineering backgrounds.

There are several technological opportunities for applications, including electric power plants; multimegawatt portable power supplies for land, air, or space uses; high-enthalpy test facilities for testing high-speed propulsion systems; magnetoplasmadynamic (MPD) thrusters for space vehicles; and MHD boost of oxygen-fuel jets in coating applications. The ultimate users of MHD power generation technology may be the utilities and independent power producers who generate electric power and who seek economic and environmentally benign methods of generating it.

Major achievements during the past 10 years include the development of equipment capable of operating for long durations and a greater understanding of the physical phenomena associated with the corrosion and erosion of plasma-facing surfaces. Industry has played a key role in the engineering development of components and subsystems of power-generating systems for proof-of-concept demonstrations. The ultimate objective of this work is commercialization of the technology.

The science community involved in this area is small. There is a need for a better understanding of basic MHD phenomena. Specific needs include a detailed understanding of conditions that lead to plasma instabilities at high power densities; a better understanding and analysis of electric discharges in flowing, reacting gases; analysis and experiments to better understand electrode and boundary layer phenomena in the presence of strong magnetic fields, slag-layer shorting effects, and associated electrical nonuniformities; and a better understanding of scaling laws. Additional research could also foster the development of high-temperature, nonslagging channel walls and high-temperature direct-fired air preheaters.

PLASMAS FOR POLLUTION CONTROL AND REDUCTION

With the increased concern for the environment, the use of plasmas for pollution control and reduction is predicted to be an area of considerable growth in the next decade. In particular, American industry is beginning to realize the importance of low-temperature plasmas for pollution control (e.g., flue gas treatment, air toxics treatment), while international efforts at pilot-plant scale are much more advanced. At present, researchers in this community are focusing mainly on studies of the plasma chemistry and discharge physics of nonequilibrium plasmas. The plasmas are usually created by electrical or electron-beam-driven discharges. This field is very old in terms of the phenomenological understanding of the phenomena involved and the identification of potential practical applications. Modern research in this field is highly applications oriented, with basic research focused primarily on the measurement and computer-based modeling of transient events. Applied research in this area is divided between a fundamental approach, involving basic discharge physics and plasma chemistry, and an Edisonian approach, centered on plasma-based pollution control devices.

Nonequilibrium plasma technology has been applied to the chemical processing of gaseous media for more than a century. Two major applications are chemical synthesis, exemplified by ozone generation, and the removal of undesirable compounds from flue gases, exemplified by the electrostatic precipitator. During the past two decades, interest in applying nonequilibrium plasmas to the removal of hazardous chemicals from gaseous media has been growing, particularly because of heightened concerns over the pollution of our environment and a growing body of environmental regulations. These more recent applications have involved efforts to destroy toxic chemical agents, to remove harmful acid rain gases such as sulfurous and nitrous oxides, and to treat other environmentally hazardous hydrocarbon and halocarbon compounds. Major contributions in the last 10 years include the decontamination of wastewater, flue/stack gas processing for SO_2 and NO_x reduction, military applications (nerve agent destruction), and toxic chemical/vapor processing.

Industry is a potential user and market for plasma technology for pollution control. The utility industry is faced with more stringent environmental regulations, which demand improved technology for effluent cleanup (both for power plants and for utility customers). Industry can play a strong role as an advocate for technology development and as a technical contributor by working cooperatively with researchers in the field on applications. Many institutions are actively involved in this area, including Lawrence Livermore National Laboratory, Los Alamos National Laboratory, Sandia National Laboratories, and several universities and industrial laboratories. To date, funding in this area has been small, but given the increasing concern about environmental issues, it could increase dramatically in the next decade.

There is much that has to be learned before low-temperature plasmas can be used in the cleanup and preservation of the environment. This includes developing a better database for plasma chemical processes, reaction-rate constants, and the resulting products, and developing diagnostics to determine the physics and chemistry of the cleanup process. There are many scientific and technical opportunities, including developing basic plasma data on the reaction of excited states and radicals with various contaminants and sophisticated modeling of the physics and chemistry of plasmas as they apply to the cleanup problem. Fortunately, there are many facilities in the various national laboratories, universities, and industry that can be used to perform the initial proof-of-principle experiments. There is a strong synergism between environmental applications and the many other disciplines that also depend on low-temperature plasmas.

PLASMA PROCESSING OF MATERIALS

Plasmas used for the processing of materials affect several of the largest manufacturing industries, including national defense, automobiles, biomedicine, computers, waste management, paper, textiles, aerospace, and telecommunications. The importance of plasma processing to the electronics industry is illustrated in Figure 1.1. An NRC study reviewed plasma processing of materials in detail in 1991.[2] This section briefly summarizes its findings and recommendations.

Applications of plasma-based systems used to process materials are diverse because of the broad range of plasma conditions, geometries, and excitation methods that may be used. This technology is multidisciplinary and, ideally, the researchers should have a basic knowledge of several scientific disciplines, including elements of electrodynamics, atomic science, surface science, computer science, and industrial process control. The impact of—and urgent need for—plasma-based materials processing is overwhelming for the electronics industry.

In its report, the NRC study panel made the following statements:[3]

- "In recent years, the number of applications requiring plasmas in the processing of materials has increased dramatically. Plasma processing [such as that illustrated in Plate 1] is now indispensable to the fabrication of electronic components and is widely used in the aerospace and other industries. However, the United States is seeing a serious decline in plasma reactor development that is critical to plasma processing steps in the manufacture of VLSI [very large scale integrated] microelectronic circuits. In the interest of the U.S. economy and national defense, renewed support for low-energy plasma science is imperative." (p. 2)
- "The demand for technology development is outstripping scientific understanding of many low-energy plasma processes. The central scientific problem underlying plasma processing concerns the interaction of low-energy collisional plasmas with solid surfaces. Understanding this problem requires knowledge and expertise drawn from plasma physics, atomic physics, condensed matter physics, chemistry, chemical engineering, electrical engineering, materials science, computer science, and computer engineering. In the absence of a coordinated approach, the diversity of the applications and of the science tends to diffuse the focus of both." (p. 2)
- "Currently, computer-based modeling and plasma simulation are inadequate for developing plasma reactors. As a result, the detailed descriptions required to guide the transfer of processes from one reactor to another or to scale

[2] See footnote 1, p. 36.
[3] See footnote 1, p. 36.

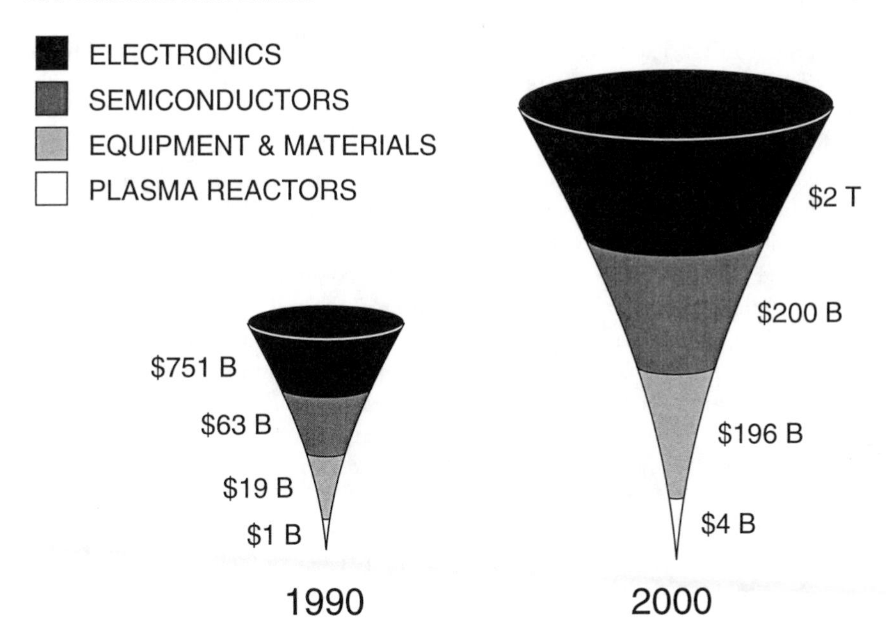

FIGURE 1.1 The world electronics "food chain." Although revenues for plasma technology are a small portion of the world electronics market, plasma technology is a critical component upon which the industry rests. Note that the plasma reactor business is expected to quadruple in this decade. (Courtesy of R.A. Gottscho. Adapted from the National Advisory Committee on Semiconductors report, *Preserving the Vital Base*, Arlington, Va., July 1990, and from data in "Semiconductor Equipment Manufacturing and Materials Worldwide," Dataquest, Inc., 1994.)

processes from a small to a large reactor are not available. Until we understand how geometry, electromagnetic design, and plasma-surface interactions affect material properties, the choice of plasma reactor for a given process will not be obvious, and costly trial-and-error methods will continue to be used. Yet there is no fundamental obstacle to improved modeling and simulation nor to the eventual creation of computer-aided design (CAD) tools for designing plasma reactors. The key missing ingredients are the following: (1) A reliable and extensive plasma data base against which the accuracy of simulations of plasmas can be compared. . . . (2) A reliable and extensive input data base for calculating plasma generation, transport, and surface interactions. . . . (3) Efficient numerical algorithms and supercomputers for simulating magnetized plasmas in three dimensions." (p. 3)

• "In the coming decade, custom-designed and custom-manufactured chips, i.e., application-specific integrated circuits (ASICs), will gain an increasing fraction of the world market in microelectronic components. This market, in turn, will belong to the flexible manufacturer who uses a common set of processes and equipment to fabricate many different circuit designs. Such flexibility in processing will result only from real understanding of processes and reactors. On the other hand, plasma processes in use today have been developed using a combination of intuition, empiricism, and statistical optimization. Although it is unlikely that detailed, quantitative, first-principles-based simulation tools will be available for process design in the near future, design aids such as expert systems, which can be used to guide engineers in selecting initial conditions from which the final process is derived, could be developed if gaps in our fundamental understanding of plasma chemistry were filled." (p. 4)

• "Three areas were recognized by the PLSC [Plasma Science Committee] panel as needing concerted, coordinated experimental and theoretical research: surface processes, plasma generation and transport, and plasma-surface interactions. For surface processes, studies using well-controlled reactive beams impinging on well-characterized surfaces are essential for enhancing our understanding and developing mechanistic models. For plasma generation and transport, chemical kinetic data and diagnostic data are needed to augment the basic plasma reactor CAD tool. For studying plasma-surface interactions, there is an urgent need for in situ analytical tools that provide information on surface composition, electronic structure, and material properties." (p. 4)

• "Breakthroughs in understanding the science will be paced by development of tools for the characterization of the systems. To meet the coming demands for flexible device manufacturing, plasma processes will have to be actively and precisely controlled. But today no diagnostic techniques exist that can be used unambiguously to determine material properties related to device yield. Moreover, the parametric models needed to relate diagnostic data to process variable are also lacking." (p. 4)

• "The most serious need in undergraduate education is adequate, modern teaching laboratories. Due to the largely empirical nature of many aspects of plasma processing, proper training in the traditional scientific method, as provided in laboratory classes, is a necessary component of undergraduate education. The Instrumentation and Laboratory Improvement Program sponsored by the National Science Foundation has been partly successful in fulfilling these needs, but it is not sufficient." (p. 5)

• "Research experiences for undergraduates made available through industrial cooperative programs or internships are essential for high-quality technical education. But teachers and professors themselves must first be educated in low-energy plasma science and plasma processing before they can be expected to educate students. Industrial-university links can also help to impart a much needed, longer-term view to industrial research efforts." (p. 5)

Plasma processing of materials is a technology that is of vital importance to several of the largest manufacturing industries in the world. Foremost among these industries is the electronics industry, in which plasma-based processes are indispensable for the manufacture of VLSI microelectronic circuits (or chips). Plasma processing of materials is also a critical technology in the aerospace, automotive, steel, biomedical, and toxic waste management industries. Because plasma processing is an integral part of the infrastructure of so many American industries, it is important for both the economy and the national security that the United States maintain a strong leadership role in this technology.

As in the case of other disciplines that use low-temperature plasmas, there is no centralized agency that takes responsibility for R&D for this area. The NRC plasma processing study determined that there are approximately 14 agencies within the federal government that invest approximately $17 million in plasma process science and technology. It concluded that this funding was inadequate and uncoordinated, given the impact of this vital area on the country.

CONCLUSIONS AND RECOMMENDATIONS

Conclusions

Low-temperature plasma science has significantly improved the quality of our lives. These contributions will continue by providing solutions to several present and future problems and by preserving our industrial base and providing challenging opportunities in the post-Cold War era. Examples of technical areas that will benefit from low-temperature plasmas include the plasma processing of materials, environmental cleanup, "cold" sterilization of medical products, "cold" pasteurization of food, advanced imaging devices that can be used in medicine and in the detection of explosives and drugs, and isotope separation.

Research in low-temperature plasmas has decreased substantially, primarily because the largest source of funding, the federal government, has had a shrinking budget for such activities in the last several years. Research has also been adversely affected by the recent recession and a general move of large U.S. companies to divest themselves of manufacturing.

The shrinking budgets of the last few years have resulted in a sharp decrease in the population of scientists working in low-temperature plasma science. The supply of PhD-level scientists would be sufficient to reverse this trend, if funding were available. If this trend is not reversed, the United States will be creating a future problem.

Recommendations

To fully exploit the potential of low-temperature plasma science and maximize its impact on the many relevant technological applications, the panel recommends that one agency within the government be given the responsibility for coordinating research in low-temperature plasma science. Given the multidisciplinary nature of the field, the Advanced Research Projects Agency (ARPA) and the National Institute of Standards and Technology (NIST) are possible candidate agencies for this responsibility.

The panel recommends that the responsible agency focus on funding low-temperature plasma science. Such projects would include the physics and chemistry of the plasma sheath; plasma stability; electrodeless plasma production; magnetic field effects on plasmas; improved diagnostics to help understand surface and sheath effects and plasma stability; energy and charge transfer from plasmas to particulates; and improved utilization of computers.

2

❖

Nonneutral Plasmas

INTRODUCTION AND BACKGROUND

A nonneutral plasma is a many-body collection of charged particles in which there is not overall charge neutrality. Such systems are characterized by self-electric fields and, in high-current configurations, self-generated magnetic fields. *Single-component plasmas* are an important class of nonneutral plasmas, the most common examples of which are pure electron and pure ion plasmas. For single-component plasmas in cylindrical geometry, there exists a stringent confinement theorem. The practical consequence of this theorem is that, in contrast to electrically neutral plasmas, a magnetized single-component plasma can be confined easily for very long times (e.g., hours). Therefore, thermal equilibrium and controlled departures from equilibrium can be achieved readily. Nonneutral plasmas exhibit a broad range of collective plasma behavior, such as plasma waves, instabilities, and Debye shielding. Moreover, the rotation and self-generated fields in these plasmas can have a significant effect on plasma properties and stability behavior.

In addition to their importance in understanding fundamental aspects of the behavior of many-body charged-particle systems, there are many practical applications of nonneutral plasmas. Examples discussed elsewhere in this report include the generation of coherent radiation by intense charged-particle beams, the development of advanced accelerator concepts, and the stability of electron and ion flow in high-voltage diodes. Other applications include particle-beam fusion, and the stability and propagation of intense charged-particle beams through background plasma or through the atmosphere. This section focuses specifically on single-component plasmas, with emphasis on pure electron and pure ion plasmas. Research and development in this area is likely to have a significant impact on a wide range of important applications, such as new generations of precision clocks; chemical analysis by improved methods of mass spectrometry; and the accumulation, storage, and transportation of antimatter.

Early research on nonneutral plasmas predated by many decades common usage of the term "plasma." For example, efforts to investigate the equilibrium and stability properties of nonneutral electron flow began with Child (1911), and continued with the work of Langmuir (1923), Llewellyn (1941) and Brillouin (1945), and work on beam-type microwave devices in the 1940s and 1950s. During the past 20 years, interest in the physics of single-component plasmas has grown substantially in such diverse areas as the equilibrium, stability, and transport properties of these plasmas; phase transitions in two- and three-dimensional plasmas; astrophysical studies of large-scale nonneutral plasma regions in the magnetospheres of neutron stars; and the development of positron and antiproton ion sources.

In the case of trapped-ion plasmas, an important synergism has developed between atomic physicists and plasma physicists. Atomic physicists have developed methods to confine and study small collections of ions with great precision. With the addition of more particles, issues of collective oscillations and confinement properties of spatially extended, three-dimensional plasmas become relevant and raise a number of important questions. Study of these questions has illuminated fundamental issues in plasma physics. It has resulted in enhanced capabilities in the creation and control of pure ion plasmas for precision measurements of fundamental constants and for applications such as atomic clocks.

A significant fraction of nonneutral plasma research is closely tied to important technological applications. In contrast to the general development of fundamental plasma experiments that has been hindered significantly in the past two decades due to lack of support, experimental progress in nonneutral plasma research has been excellent, which has stimulated much progress in the theory of nonneutral plasmas. It is the conclusion of the panel that the relative success of research on nonneutral plasmas was due to a strong and dedicated program of support in this area by the Office of Naval Research, with complementary support from the National Science Foundation and the Department of Energy. The panel concludes that this mode of support for nonneutral plasma research should be considered a model for the support of fundamental plasma experiments in the broader area of neutral plasma research.

RECENT ADVANCES IN NONNEUTRAL PLASMAS

The following summarizes significant advances in the physics of nonneutral plasmas during the past decade.

Electron Plasmas

Much progress has been made in understanding the basic physics of single-component plasmas, including critical aspects of the stability, confinement, and equilibrium of these plasmas. It was shown theoretically that the conservation of canonical angular momentum implies, in the absence of external torques, that a single-component plasma can be confined indefinitely. Soon afterward, it was demonstrated that the confinement of pure electron plasmas for several minutes to hours is relatively easily achievable in laboratory experiments. The confinement times are sufficiently long that the plasma approaches a state of thermal equilibrium.

The existence of these thermal equilibrium states in confined, single-component plasmas distinguishes them from neutral plasmas. A magnetically confined neutral plasma does not remain in a state of spatially isolated, local thermal equilibrium, because collisions between the electrons and ions lead to a diffusive expansion of the plasma across magnetic field lines. In addition, in a neutral plasma there is typically free energy (associated with the relative cross-field flow of electrons and ions) available to drive collective instabilities that produce enhanced transport across the field lines. Such instabilities pose a challenge to the achievement of high-quality confinement in electrically neutral plasmas of interest in fusion. In contrast, a confined, single-component plasma that has come to thermal equilibrium is in a state of minimum free energy and hence is stable. It is also a great advantage theoretically to be able to use thermal equilibrium statistical mechanics to describe the equilibrium state.

Theory predicted that, in a strong magnetic field and at low temperature, the relaxation of the particle velocities to a thermal equilibrium distribution would be constrained by an adiabatic invariant, and as a consequence, the relaxation rate would be exponentially small. Subsequent experiments confirmed this prediction, and now there is good agreement between theory and experiment over eight orders of magnitude in effective magnetic field strength and five orders of magnitude in the scaled relaxation rate.

The well-controlled nature of these plasmas has also permitted precise studies of nonequilibrium states unachievable in other plasmas. For a sufficiently low-density nonneutral plasma, in the limit that transport along magnetic field lines is rapid compared to transport perpendicular to the field, the plasma is described by similar equations (in an isomorphic sense) to those describing an inviscid classical fluid in two dimensions. Charge-density perturbations in a single-component plasma are analogous to vortices in a fluid, and vortex dynamics is an important subject of long-standing interest in fluid dynamics. Recently, this analogy has begun to be exploited to test models of coherent structures and vortex merger with a precision not possible in classical fluids. For example, since the effective viscosity of a pure electron plasma is less by orders of magnitude than the viscosity of a classical fluid, the trajectories of a pair of vortices can be followed for 10^4 or 10^5 orbits before merger occurs. In the case of classical fluids such as water, merger or dissipation typically occurs in a few orbits.

Finally, a nonneutral plasma exhibits a wide range of collective waves and instabilities analogous to those observed in an electrically neutral plasma, appropriately modified by self-field effects. These collective waves and instabilities have been documented extensively in theoretical analyses, and algorithms have been developed for calculating the detailed stability behavior of nonneutral plasma over a wide range of system parameters. In addition, a kinetic stability theorem has been developed that determines a sufficient condition for the nonlinear stability of cylindrically symmetric equilibria to arbitrary-amplitude perturbations, including the influence of strong self-field effects.

Ion Plasmas

Another type of single-component plasma that has been studied extensively is the magnetized, pure ion plasma, confined in a Penning trap, and cooled by laser radiation. In this case, the laser light is used both to cool the ions and to exert a torque on the plasma. This torque has the effect of spinning up the plasma and compressing it. An analytical theory of the collective modes of oscillation in these plasmas has been formulated on the basis of cold fluid theory. This is the first analytical description of the modes of a magnetized, three-dimensional plasma of finite extent with realistic boundary conditions. There is good agreement between theory and the experimental observation of these modes. One result of this increased understanding is that these modes can now be used for the manipulation and confinement of pure ion plasmas. An exact nonlinear theory has been developed for the case of large-amplitude quadrupole modes of oscillation of these plasmas.

In ion plasmas that are laser-cooled to cryogenic temperatures, the average kinetic energy per particle can be made small compared to the average interaction potential energy. (These plasmas are often referred to as strongly coupled plasmas.) The resulting ion clouds can form the analogues of dense liquid and solid phases. (See Figure 2.1.) Theoretical and experimental progress has been made recently in understanding the ordering and equilibrium states of these systems. As the temperature is lowered, theory predicts that the ions arrange themselves in concentric spheroidal shells. These shells are the analogues of crystal planes, except that the planes are deformed into spheroids because of the small plasma size. This shell structure has been observed experimentally, with optical imaging techniques, for plasmas up to about 15 shells. Theory predicts that the sample must contain about 60 shells to result in the structure predicted for plasmas of infinite extent (a body-centered-cubic lattice), but this has not yet been tested experimentally.

Small numbers of ions have also been confined and cooled in Paul traps, which utilize the ponderomotive force from high-frequency electric fields to confine the ions. At low temperature, Coulomb repulsion between the ions causes the ions to crystallize into simple geometrical configurations (Coulomb clusters) whose shapes can be predicted theoretically. These clusters and ordered one-dimensional chains of ions have now been observed and studied in Paul traps. (See figure 2.1a.) As the number of ions is increased, experiments have observed polymorphic phase transitions to more complex lattice structures: first a zigzag arrangement, then a helical chain, and finally a cylindrical shell structure, similar to the spheroidal shells observed in Penning traps. These phase transitions have also been studied theoretically.

Such linear lattice structures also are predicted to occur in chains of ions confined and cooled in a heavy-ion storage ring. In the rest frame of the ions circulating in such a storage ring, the confining forces are nearly the same as those in the linear Paul trap described above.

The theoretical density limit for the confinement of a magnetized, single-component plasma occurs when the square of the plasma frequency is one-half the square of the cyclotron frequency (Brillouin, in 1945). This "Brillouin density limit" has been achieved in pure ion plasmas by using laser radiation to exert torques on the plasma and thereby to compress it. In a "cold" one-component plasma column, the radially outward space-charge and centrifugal forces on a fluid element balance the inward magnetic confining force (i.e., the Lorentz force). This places a limit on the maximum plasma density that can be confined for a given value of magnetic field (i.e., the Brillouin density limit). The rotation frequency at the Brillouin limit is such that the Lorentz force on the plasma particles is just canceled by the Coriolis force, and the plasma is effectively unmagnetized when viewed in the rotating frame. Therefore, at the Brillouin limit, it is possible to study in detail a fundamentally new plasma regime in which the confined plasma is effectively "unmagnetized."

Ion Plasmas in Electron-Beam Ion Traps

The electron-beam ion trap configuration, which was invented in the last decade, uses a magnetically compressed electron beam, with energies in the range of several hundred keV, to ionize, trap, and excite highly charged ions of a wide variety of elements for atomic physics measurements. Electron-beam ion trap devices are capable of producing high-resolution x-ray spectra of nearly stationary ions that have been excited by monoenergetic electrons. One can also vary the energy of the electron beam on a time scale fast compared to that for changes in the ionization states of the ions. Thus, the ions can be excited with electrons of one energy and probed with electrons of a different energy.

These devices have been able to produce one-electron (i.e., hydrogen-like) ions up to nuclear charge $Z = 92$. In the last few years many important measurements have been made utilizing electron beam ion trap devices for atomic phys-

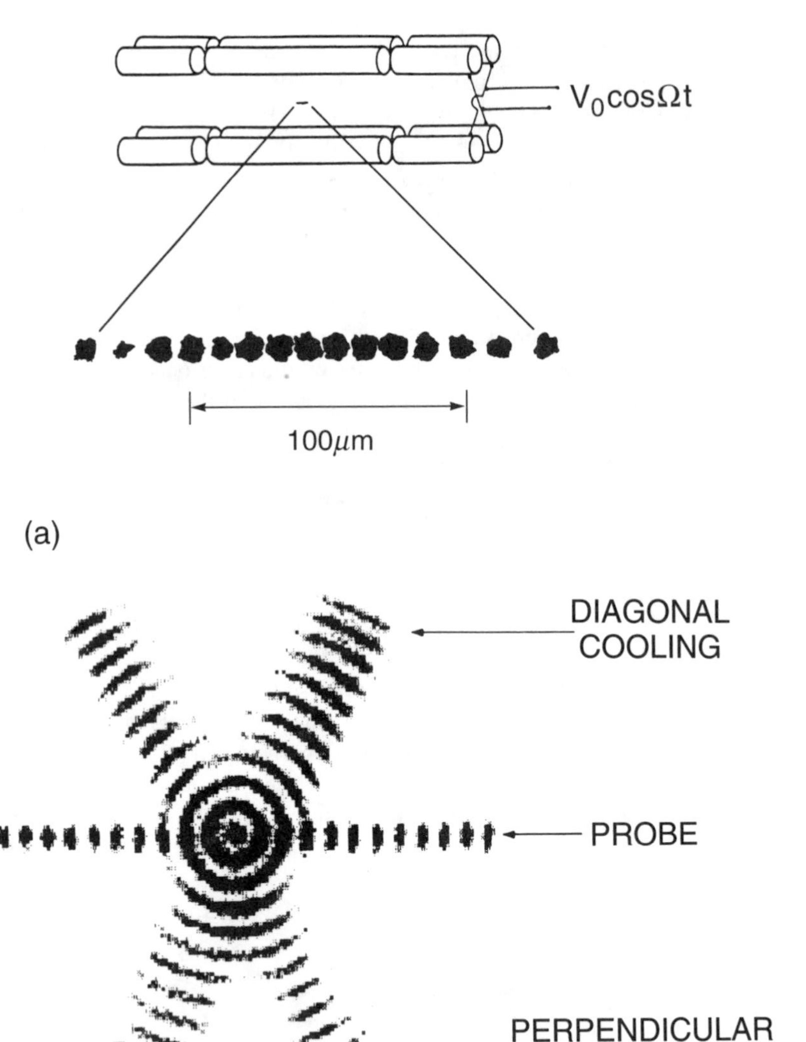

FIGURE 2.1 Correlated behavior observed when small, single-component ion plasmas are laser cooled to temperatures of a few tens of millikelvin. Such laser-cooled ion plasmas are being used to improve the performance of atomic clocks and frequency standards. (a) A crystallized chain of 15 Hg^+ ions, confined in an rf trap by the electrode structure shown. (b) Be^+ ions confined in a Penning trap, imaged by passing three crossed laser beams through the plasma. The bright fringes are the intersections of the laser

ics, including tests of leading theories of atomic structure and crucial tests of extrapolations of previous, lower-Z measurements. Other important experiments include x-ray observations of magnetic octopole decay in atomic spectra; the first use of x-ray polarization to probe the hyperfine interaction in highly charged ions; the first direct measurements of ionization cross sections for highly charged ions; the first measurements of dielectronic recombination cross sections in ions that are important in hot fusion plasmas; the first excitation functions for x-ray lines used in the analysis of high-temperature tokamak and astrophysical plasmas; measurements of line overlaps for x-ray laser design; and measurements of metastable lifetimes in regions of the electromagnetic spectrum inaccessible to other techniques.

Confinement of Antimatter

Trapping techniques similar to those described above for pure electron plasmas have proved to be an efficient way to accumulate and store antimatter particles such as positrons and antiprotons. Single-component positron plasmas, a few cubic centimeters in volume, with a temperature of 300 K and Debye screening lengths less than 1 mm have now been created in the laboratory by accumulating positrons from a radioactive source. Recently, antiprotons from the low-energy antiproton storage ring at CERN in Geneva have been captured by a similar trapping scheme. The antiprotons were cooled to 4 K by collisions with an electron plasma confined in the same cryogenic trap.

These experiments have provided a controlled way to study antimatter interactions with ordinary matter and to study the properties of the antimatter particles themselves. Examples include precision measurements of the mass of the antiproton and positron annihilation phenomena relevant to atomic and molecular physics and to gamma-ray astronomy.

RESEARCH OPPORTUNITIES

Continued progress is expected in the areas identified above in "Recent Advances in Nonneutral Plasmas." In addition, the following topics, while not comprehensive or mutually exclusive, represent important research opportunities in the physics of nonneutral plasmas.

---

beams with the plasma's lattice planes, which take the form of approximately spheroidal shells. The plasma rotates about its symmetry axis (normal to the figure), which obscures the image of individual ions within each shell. (Courtesy of J. Bollinger and D. Wineland, National Institute of Standards and Technology, Boulder, Colo.)

Coherent Structures and Vortex Dynamics

The progress made during the past decade in studies of single-component plasmas has created a number of important scientific and technological opportunities. The ability to confine and manipulate pure electron plasmas and to create near-equilibrium states opens up unique opportunities to study transport in plasmas and fluids. The equations that govern two-dimensional flows in these plasmas are identical to the equations that govern two-dimensional flows in inviscid incompressible fluids. Exploiting this analogy, researchers have now conducted sucessful studies of vortex merger in electron plasmas, and the opportunity now exists to study important phenomena in fluid mechanics, such as the relaxation of two-dimensional turbulence (see Plate 2), the interaction of vortices with turbulence and shear flows, and turbulent transport.

Transport Processes

On longer time scales, the transport of particles due to like-particle collisions is only partially understood. In principle, this is a more difficult problem than the interaction of two-dimensional vortices because three-dimensional effects may be important, as well as the combined effects of single-particle and collective interactions. These effects are related to fundamental issues in kinetic theory and transport processes in neutral plasmas, but they can be isolated and studied more easily in single-component plasmas because of the unusually long confinement times.

Confinement Properties in Nonaxisymmetric Geometries

Recently, magnetized, single-component electron plasmas have been created that are *not* symmetric in the plane perpendicular to the confining magnetic field. (See Figure 2.2.) These plasmas were found to have surprisingly long confinement times. It does not appear that these long-lived, asymmetric states can be explained by the simplest models of good confinement of single-component plasmas with cylindrical symmetry, which indicates that the fundamental principles of single-component plasma confinement are not fully understood.

Stochastic Effects

In complex magnetic field geometries, the combined influence of the applied field configuration and the self-electric and self-magnetic fields of the nonneutral electron or ion beam can significantly affect individual particle motion and beam dynamics. For example, this can occur in the periodic wiggler field in free-electron lasers or in the periodic quadrupole focusing field in induction accelerators, particularly at sufficiently high beam intensities. Although the

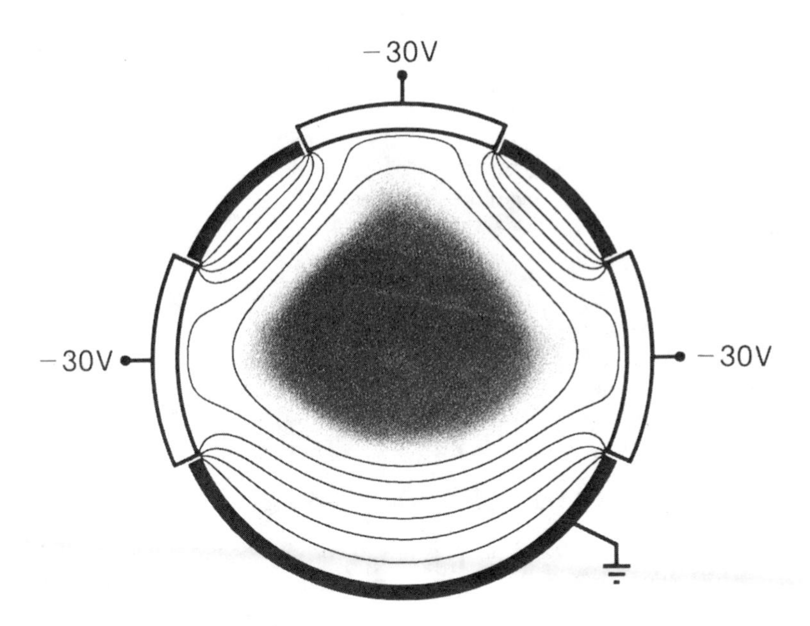

FIGURE 2.2 Shown is the cross section of a magnetized pure-electron plasma confined in a Penning trap. These plasmas were found to exhibit unexpectedly long confinement times. This good confinement is consistent with a recently developed theory that argues that such states are stable equilibria. The plasma is distorted into a triangular shape by the application of electrical potentials (indicated in volts) to sections of a cylindrical electrode structure. The calculated equipotential contours (solid lines) illustrate that the plasma edge follows such a contour. Note that the electrons are closer to the negative electrodes, as expected for a state of maximum electrostatic energy, and as predicted by the theory. (Reprinted, by permission, from J. Notte, A.J. Peurrung, J. Fajans, R. Chu, and J.S. Wurtele, Physical Review Letters 69:3056, 1992. Copyright © 1992 by the American Physical Society.)

particle dynamics are characterized by well-defined single-particle constants of the motion at low beam intensity, where self-field effects are negligibly small, at higher beam intensity the particle orbits can become chaotic and sensitive to the detailed properties of the beam density and current profiles. We do not have a basic understanding of the influence of stochastic effects on the charge homogenization in periodic focusing quadrupole configurations or on the suppression of coherent free-electron-laser emission at high beam intensity.

Strongly Coupled Nonneutral Plasmas

Strongly coupled pure ion plasmas present another set of scientific opportunities. The evolution of the spatial ordering to the body-centered-cubic structure that is expected for a large-volume plasma remains to be studied experimentally. Such experiments will test the predictions of recent theories. Furthermore, for modest-sized plasmas, the theoretical prediction of ordering in concentric spheroidal shells is at variance with observations of open cylindrical shells. This significant discrepancy also remains to be reconciled.

Although there is now a good understanding of many features of the equilibria of ion plasmas and of the linear dynamics of these plasmas about their equilibria, many important questions remain. For example, the transport coefficients (such as the viscosity and the thermal relaxation time) in strongly correlated, magnetized, nonneutral plasma are not understood theoretically. Experiments can measure many of these coefficients. It is likely that future interplay between theory and experiment in this area will be productive in elucidating these fundamental transport processes.

Quantum-Mechanical Effects

The tools now exist to create plasmas in correlated spin states and to create quantized plasmas, in which the quantum-mechanical ground state energy is large compared to the thermal energy in a plasma mode. Another important area for future study relates to the properties of nonneutral plasmas at or near the Brillouin density limit.

Antimatter

The efficient trapping of single-species plasmas has been exploited for the confinement and cooling of both antiprotons and positrons. Further progress in developing efficient techniques for the confinement and manipulation of single-species plasmas will directly benefit studies of antimatter and the interaction of antimatter with ordinary matter. Improvements in the trapping and manipulation of single-component plasmas will lead to the ability to transport antimatter (such as antiprotons) from high-energy accelerators, where they are created, to laboratories throughout the world.

Many important scientific questions can be addressed by collections of antimatter particles confined in traps. For example, one can study the physics of electron-positron plasmas. These plasmas are unusual in the sense that both signs of charge can be highly magnetized, and the "electron-ion" mass ratio is unity. Important physics issues include the nature of confinement and transport in these neutral but highly magnetized plasmas and the nature of fluctuations and turbulence in such equal-mass plasmas. The interaction of low-energy positrons with ordinary matter can also be studied with precision in traps, to address questions relevant to atomic and molecular physics and to gamma-ray astronomy. The 511-keV gamma-ray annihilation line is the strongest astrophysical source of gamma-ray line radiation. Trapped antiprotons will be of use for fundamental physics measurements on the antiprotons themselves. Moreover, addition of a cold positron plasma to the trapped antiprotons is thought to be the simplest way to produce cold antihydrogen, which would be the first creation of neutral antimatter in the laboratory.

OPPORTUNITIES FOR ADVANCES IN TECHNOLOGY

Precision Clocks

A wide range of technological opportunities is likely to result from research on laser-cooled ion plasmas. Spectroscopic interrogation of ions in traps is one of a few possible techniques for developing a new generation of precision clocks. Improvements in clocks using pure ion plasmas hinge on understanding the microscopic distribution function of the ions or, equivalently, on a more quantitative theoretical understanding of strongly coupled plasmas. Improved precision clocks would lead to advances in such diverse areas as navigation and tests of general relativity.

Precision Mass Spectrometry

One of the most important techniques for studying the masses of chemical species is ion cyclotron resonance, where the cyclotron motion of a confined cloud of ions is excited and detected. A large signal-to-noise ratio requires a large number of ions, but in this case precise interpretation of the cyclotron resonance signal hinges on a detailed understanding of the collective modes of oscillation of these multispecies ion plasmas. Scientific issues in this area have only recently begun to be addressed, starting with studies of the cyclotron modes of a single-component electron and ion plasmas, and precision studies of cyclotron resonance for one or a few ions. It is likely that much progress can be made in this area during the next decade.

Ion Sources with Enhanced Brightness

Laser cooling of 1-MeV ions in storage rings has recently been achieved. The development of methods for cooling these single-component plasmas and understanding their behavior can lead to brighter ion beams and hence to enhanced accelerator performance. It is possible that the "sympathetic cooling" of ions confined in a trap with positrons will lead to brighter sources of positrons for advanced accelerators. The successful achievement of cryogenic plasmas opens up the possibility of preparing spin-polarized plasmas. In principle, these plasmas could provide bright sources of polarized particles for use in particle accelerators.

Electron-Beam Ion Traps

Studies of electron-beam ion traps as plasma devices are in their infancy, and progress in this area could lead to performance enhancements by several orders of magnitude. This would enable new kinds of experiments in atomic, nuclear, and surface physics. An enhanced ion source based on the electron-beam ion trap is expected to be useful for surface modification and nanotechnology. In most cases, plasma physics issues are the key to these developments. For example, a thousandfold increase in the x-ray emission rate from an electron-beam ion trap might be achieved by increasing the total electron-beam current, the current density, and the space charge neutralization (i.e., ion density). The total beam current is likely to be limited by instabilities such as those that occur in backward-wave oscillators. The current density is likely to be limited by the brightness of future electron guns, and the (poorly understood) super-emissive, hollow-cathode discharge is a leading candidate for an electron gun. The ion density will be limited by a two-stream instability. The performance of electron-beam ion traps is also limited by discharges and instabilities involving trapped secondary electrons. These phenomena are not understood to the degree necessary to design a reliable next-generation device. Progress has been made only by trial and error. It is possible that the ion output could be enhanced by an even larger amount simply by making the trap longer, but success will again depend on understanding plasma properties of these devices.

A plasma research program in this area might also spin off benefits for other electron-beam devices (e.g., klystrons, traveling-wave tubes, and free-electron lasers) and for other plasma devices (e.g., electron-cyclotron-resonance ion sources and the pure electron or pure ion plasmas described above). In addition to issues related to the electron beam itself, it is known that many electron-beam devices are affected by trapped ions. New types of devices could also evolve from the present experimental configurations of electron-beam ion traps, which, for example, might provide new and inexpensive laboratory sources of x-rays and highly charged ions and microwave devices with trapped ions designed into their operation.

Radiation Sources

The increased understanding of single-component plasmas is likely to have significant impact on the development of beam-type microwave devices, particularly for use in high-power and high-frequency applications. Such applications are discussed in more detail in the section on beams and radiation sources.

Pressure Standard in Ultrahigh-Vacuum Regime

A pure electron plasma confined in a Penning trap can potentially be used to develop a primary pressure standard in the ultrahigh-vacuum regime ($< 10^{-5}$ Pa).

The trapped electrons relax to a well-defined equilibrium state in which the average rotation frequency of the electron plasma is independent of radius. When a neutral gas is present, collisions between electrons and neutrals perturb the plasma and modify the rotation frequency and electron distribution function. By using a reference value for the elastic momentum transfer cross section of the neutrals, the neutral gas density consistent with the observed evolution of the electron plasma can be determined and used to develop a pressure standard.

SUMMARY, CONCLUSIONS, AND RECOMMENDATIONS

In the past two decades, much progress has been made in the understanding of nonneutral and single-component plasmas. New experimental configurations have been discovered and exploited, leading to a better understanding of the underlying physical principles of plasma confinement, approach to equilibrium, and in some cases, mechanisms of plasma transport. There are many potential scientific and technological uses of such plasmas. These opportunities result, at least in part, from the excellent confinement properties that distinguish single-component plasmas from neutral plasmas and enable true thermal equilibrium states to be achieved. Therefore, plasmas with controlled departures from equilibrium also can be created. This allows a study of nonequilibrium plasma phenomena with a degree of precision unachievable in other plasma systems.

Experiments in nonneutral plasmas, such as those described above, can be exploited to address forefront problems in atomic, molecular, and optical physics and in fluid dynamics, as well as in plasma physics. Consequently, it is expected that this will continue to be a vital and productive area in plasma physics research for the foreseeable future. Since these experiments can typically be done with a relatively modest expenditure of resources, they are ideally suited to a university setting.

In addition to the intrinsic scientific value of research in nonneutral plasmas, there are many important applications of these plasmas. Several examples, discussed above and in Chapter 5, "Beams, Accelerators, and Coherent Radiation Sources," include beam-type microwave devices, such as gyrotrons and free-electron lasers, precision clocks and mass spectrometers, and future generations of ion sources.

The progress in this area has benefited greatly by steady support from a dedicated program at the Office of Naval Research and support from the National Science Foundation and the Department of Energy. It is the conclusion of the panel that research on nonneutral plasmas should be considered a vital part of a healthy and vigorous plasma science program in the United States in the next decade. Therefore, the panel recommends that continued strong support be given to research on nonneutral plasmas and to the development of technological applications.

3

❖

Inertial Confinement Fusion

INTRODUCTION AND BACKGROUND

The goal of fusion research is to develop a reliable alternative to the present burning of fossil fuels for energy. In the inertial confinement approach to fusion, high-intensity laser or charged-particle beams are used to compress and heat the fusion fuel to the density and temperature required for fusion of the nuclei. The fusion of deuterium and tritium is schematically illustrated in Figure 3.1.

The pursuit of inertial confinement fusion (ICF) depends on many phenomena associated with plasma science. The interaction of radiation with matter in the plasma state and the subsequent energy transport and high-density compression leading to thermonuclear burning of the plasma fuel must be optimally balanced. Nonlinear collective effects must be understood and accommodated. The international goal is to achieve an environmentally improved source of electrical power generation. The majority of the program continues to be implemented in the nuclear weapons laboratories, the Naval Research Laboratory, and the Laboratory for Laser Energetics at the University of Rochester. New large-scale facilities and facility upgrades are currently envisioned with funding authorizations at various stages. Support for the underlying basic plasma science and the breadth of the involved community should be strengthened. The role of basic plasma research within the ICF program may be at a crossroads, requiring timely reexamination.

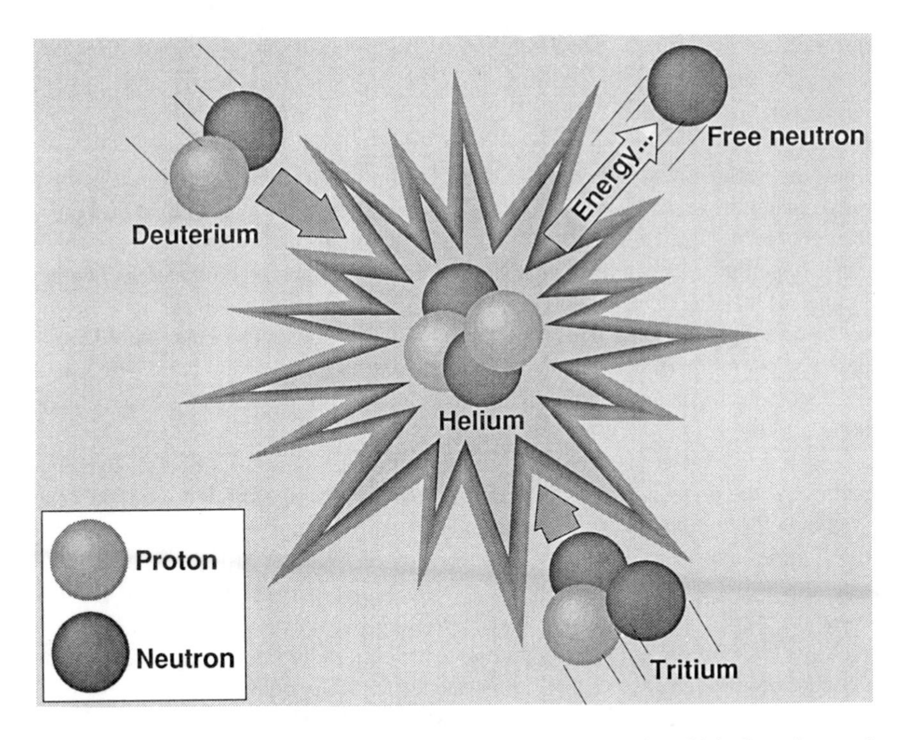

FIGURE 3.1 Schematic diagram of the basic fusion process in which deuterium and tritium nuclei combine to form ^{3}He and a neutron. For electric power applications, the energy from this reaction is transformed into heat and then converted into electrical energy.

RECENT ADVANCES

Laser Fusion

In the past decade, significant progress has been made in the understanding of high-energy-density plasmas created by intense lasers and particle beams. An ICF hohlraum irradiated by the Nova laser is shown in Plate 3. With several notable exceptions, this work has been carried out under the auspices of inertial confinement fusion research with large lasers (i.e., having energies greater than 1 kJ). Direction, progress, and accomplishments within the ICF program have been subject to frequent national review.[1]

Experiments and computer simulations during the past decade have led to a

[1]For example: National Research Council, *Second Review of the Department of Energy's Inertial Confinement Fusion Program*, Final Report, National Academy Press, Washington, D.C., 1990.

quantitative understanding of the Rayleigh-Taylor instability in hot, ablating plasmas. Future work will examine the transition to turbulence and address Rickmyer-Meshkov and Kelvin-Helmholtz-like instabilities. Two-dimensional hydrodynamic simulations have modeled successfully the linear and early nonlinear evolution of the Rayleigh-Taylor instability in ablating plasmas with a range of initial sources, Attwood numbers, and accelerations. Photon and electron energy deposition lead to finite density gradients and mass removal, which can substantially reduce the Rayleigh-Taylor growth rate from its classical value. Fokker-Planck codes embedded in hydrodynamic simulations have been developed to model better the nonlocal electron energy transport. These simulations describe phenomena such as thermal filamentation and thermal conduction, and influence our basic understanding of the details of the Rayleigh-Taylor instability.

Using the Nova laser facility, successful experiments were conducted that addressed the physics of interpenetrating materials at accelerated interfaces (an area of hydrodynamics critical to both ICF and weapons research). Several of the "ignition physics milestones" described in the NRC's 1990 review of the ICF program were achieved, including experimental confirmation of the LASNEX simulation code predictions for Rayleigh-Taylor instability growth rates in the presence of ablation and density gradients for both radiation-driven and electron-conduction-driven planar foils. New diagnostic techniques were demonstrated, including large neutron scintillator arrays; a single-hit scintillator array neutron spectrometer; and a high-energy, ring-aperture x-ray microscope. The Nova target chamber is shown in Figure 3.2.

There has been significant progress in the ability to measure and calculate the radiation properties of complex, partially stripped ions over a wide range of plasma conditions. The recent measurement of iron opacity in dense ($n_e > 10^{20}$ cm^{-3}), warm ($T_e \geq 70$ eV) plasmas in local thermodynamic equilibrium illustrates these advances. These conditions are also relevant for astrophysical plasmas. The demonstration of nickel-like gold plasma x-ray lasers operating at a wavelength of 33 Å is an example of the present capability to model non-LTE plasmas. Sophisticated opacity codes for dense plasmas composed of multi-electron ions have been developed. These codes describe the complex absorption and emission features of these ions in terms of unresolved transition arrays and super-transition arrays, and have led to improvements in modeling radiant energy flow in high-density plasma of interest in both inertial fusion and astrophysical applications. Laboratory experiments have been performed that validate these codes.

Ion-Beam Fusion

An equally robust and successful track record of accomplishments exists for the light- and heavy-ion ICF efforts, which represent alternative "driver" ap-

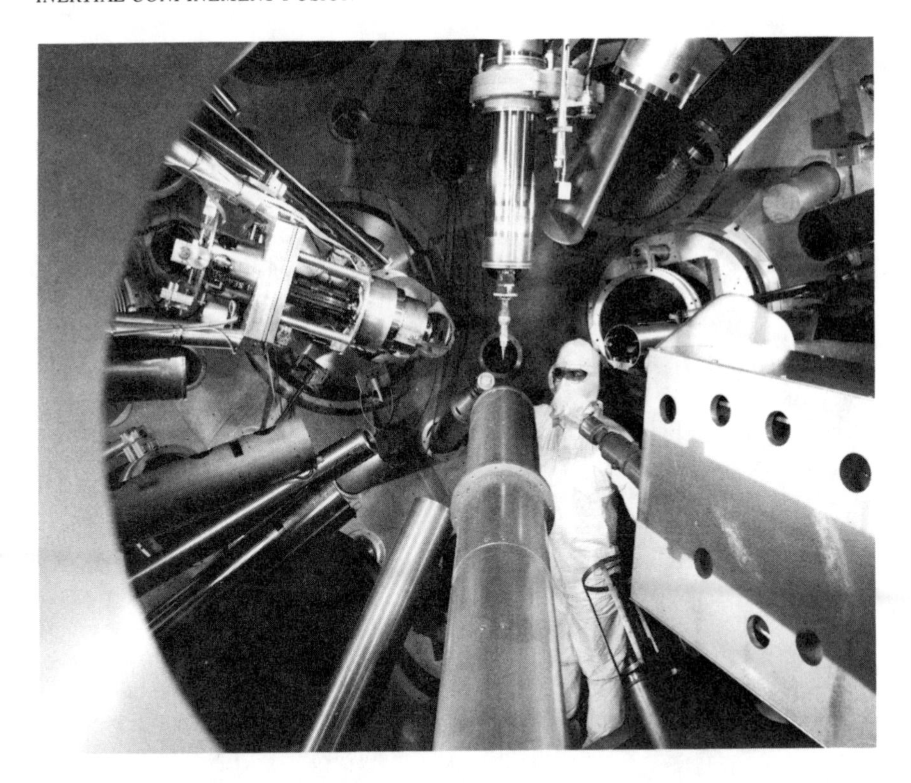

FIGURE 3.2 Photograph of the inside of the Nova 10-beam target chamber at Lawrence Livermore National Laboratory. An essential part of Nova is its diagnostic capability, which includes optical, x-ray, and neutron measurement techniques to study the performance of the ICF targets. These diagnostics surround the target, which is positioned in the center of the chamber on the end of a rod descending from the top of the picture. (Courtesy of Lawrence Livermore National Laboratory.)

proaches, and they are being pursued concurrently with the laser program. A principal aim of heavy-ion fusion accelerator research is to gain an understanding of the dynamics of intense, space-charge-dominated beams in accelerator structures. These beams, which are effectively nonneutral plasmas, exhibit collective behaviors in addition to those bulk motions driven by the externally applied fields. The beam plasma frequency is comparable to the frequency associated with motion in the applied fields. Analytic theory originally predicted a multitude of instabilities; however, most of these were not observed. Computer simulations resolved the disagreement for transverse modes by elucidating the nonlinear behavior and early saturation of the instabilities. Resolution of this disagreement represents a major step in understanding these nonneutral plasmas.

The light-ion ICF program is based on the robust and cost-effective intense ion beam technology developed over the past 20 years. The program has many significant achievements. The ability to transport megajoules of electromagnetic energy at megavolt-per-centimeter electric fields in pulse power accelerators and over several meters of transmission line has been achieved only after careful study of electron insulation in high-voltage devices. Electrons emitted from surfaces are confined by the self-magnetic field of the transverse electromagnetic (TEM) waves, and the behavior of relativistic, diamagnetic electron flow in strong electromagnetic fields is a large extrapolation beyond the physics of magnetrons. The successful generation of multi-kiloampere-per-square-centimeter light-ion beams in magnetically insulated ion diodes has resulted from careful study of the physics of collective effects in electron and ion streams in strong electric fields. This reduction in beam divergence has come from two- and three-dimensional computer codes that model the electromagnetic and particle physics in realistic geometry supported by analytical studies.

Important developments in the physics of the propagation of intense ion beams to the target in an ionized background gas have occurred both experimentally and theoretically. Unique diagnostic systems have been developed capable of measuring ion beams and plasmas on very short time scales and in extremely hostile radiation environments, including the measurement of high ion beam intensities (0.1-1.0 MA/cm^2, 1-12 MV, 1-5 TW/cm^2) using elastic scattering and characteristic x-ray line emission. Diagnostic techniques developed include visible spectroscopy, VUV spectroscopy, ion pinhole cameras, ion movie cameras, Rutherford magnetic spectrographs, and nuclear activation and nuclear track detectors with automatic track counting capability. Stark shift measurements using visible spectroscopy diagnostics on the Particle Beam Fusion Accelerator II (PBFA II) have demonstrated electric fields as high as 9 MV/cm, the largest electric field ever measured using this technique. Fluorescence spectroscopy using dye lasers has emerged as an important technique for measuring the ion-beam divergence in high-powered ion diodes.

SCIENTIFIC AND TECHNOLOGICAL OPPORTUNITIES

New facilities, such as the Omega Upgrade and the National Ignition Facility (shown in Figure 3.3), will enable the creation of plasmas with densities in excess of 10^{26} cm^{-3} and pressures exceeding 200 Gbar. Energy transport (including fusion by-products), the equation of state, and radiative properties can also be studied with these facilities. For example, plasma conditions will be well matched to the study of electron energy transport that is either nonlocal or not dominated by collisions. Configurations can also be established for magnetic fields to play a role. Dense plasmas formed by isochoric heating by penetrating electrons formed with intense short-pulse laser irradiation also may be appropriate for transport studies in which hydrodynamic expansion can be minimized.

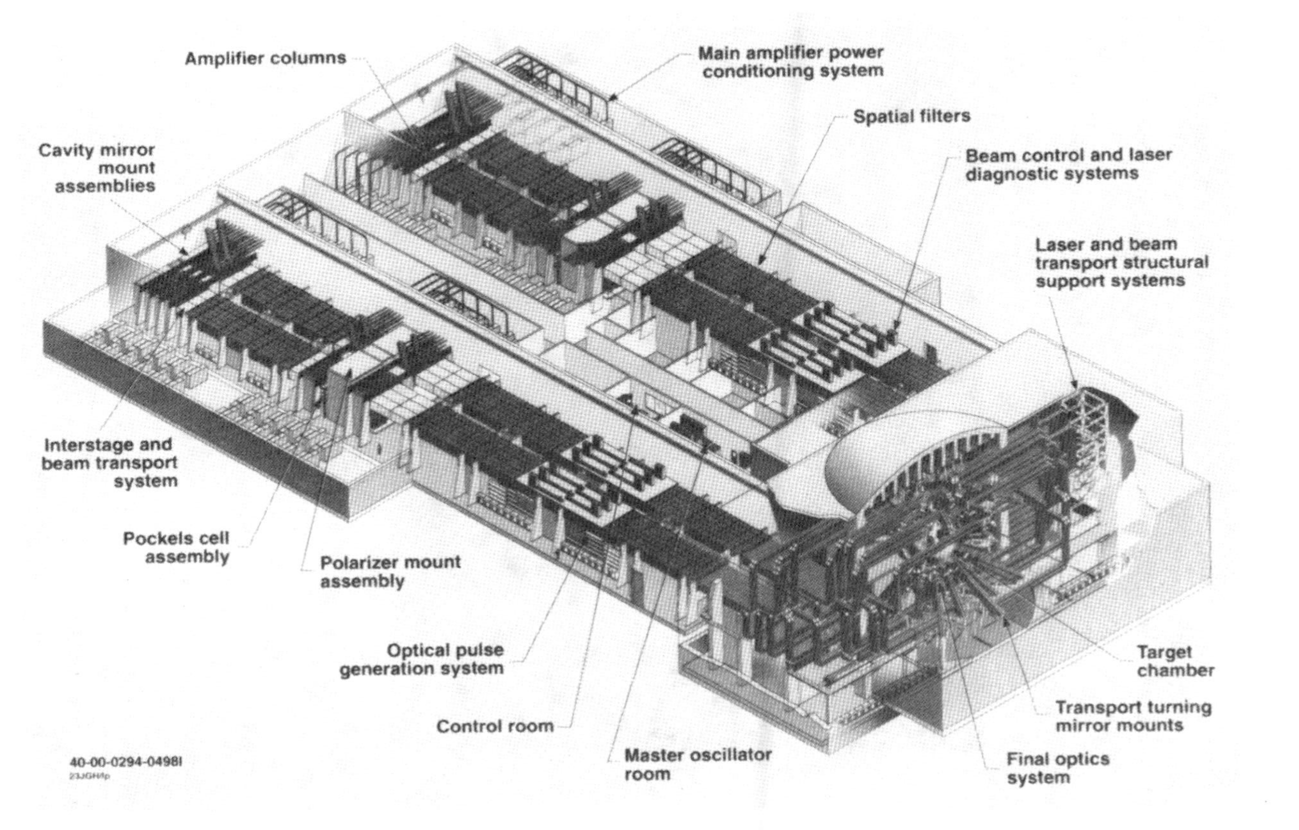

FIGURE 3.3 An artist's rendition of the proposed 192-beam National Ignition Facility (NIF). The NIF is designed to demonstrate an important ICF milestone, ignition and subsequent self-heating of the fusion fuel. Nearly 2 MJ of laser energy (40 times more than the existing Nova facility can produce) will create high-energy-density plasmas relevant to a broad range of research and applications. (Courtesy of Lawrence Livermore National Laboratory.)

These facilities will also enable quantitative experiments that examine hydrodynamic stability properties. Mixing of materials originally separated by interfaces, including the fully turbulent phase and subsequent energy, mass, and momentum transport, will be studied. The development of three-dimensional codes that incorporate turbulent mix models will complement these experiments. A laser-plasma experiment illustrating several relevant multidisciplinary phenomena is illustrated in Figure 3.4.

There will continue to be significant advances in numerical simulations and

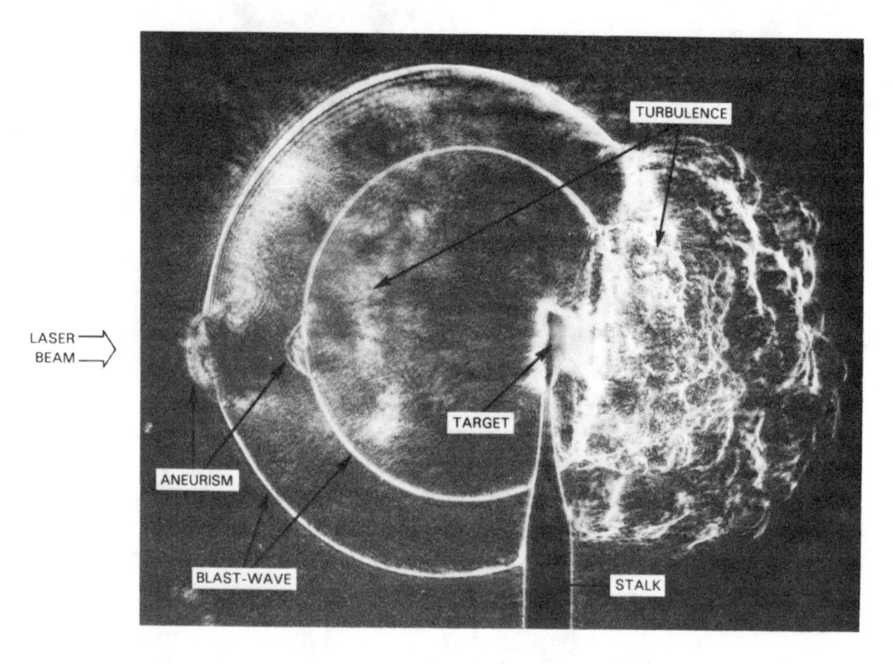

FIGURE 3.4 Images of a laser-produced plasma expanding into a gas. An intense, short-pulse laser (pulse duration 3 ns, wavelength 1.05 μm, irradiance 5×10^{13} W/cm^2) strikes a small target (an aluminum disk 1 mm in diameter and 10 μm thick, mounted on a stalk), creating an energetic, expanding plasma. The plasma initially moves to the left, and accelerated bulk target material propagates to the right. Two superimposed dark-field shadowgrams, which are sensitive to density gradients, image the evolution of a very strong shock wave (Mach number greater than 100) propagating into atmospheric pressure (left side of target stalk) and into a turbulent plasma (right side of target stalk). Such techniques permit physical effects that occur in plasmas with ultrahigh energy densities, such as space plasmas, supernovae, and nuclear explosions, to be readily studied in the laboratory. (Reprinted, by permission, from B.H. Ripin, J. Grun, C.K. Manka, J. Resnick, and H.R. Burris, "Space Physics in the Laboratory," pp. 449-463 in *Nonlinear Space Plasma Physics*, ed. R.Z. Sagdeev, AIP Press, New York, 1993. Copyright © 1993 by the American Institute of Physics.)

the experimental capability to describe the radiative properties of complex, high-energy-density plasmas. The development of non-LTE codes and expansion of the experimentally accessible parameter space will be areas of future research emphasis.

Within the heavy-ion fusion accelerator research area, additional work is needed on longitudinal modes and on the interaction between longitudinal and transverse motions. Because some processes are inherently three-dimensional, advanced three-dimensional simulations need to be developed and applied. The light-ion fusion program provides science opportunities for creating high-density and high-temperature plasmas for materials and plasma science, for developing new computer codes to investigate ion-beam transport in vacuum and plasmas, and for developing a new class of soft x-ray diagnostics that can operate in harsh radiation environments to provide accurate measurements of both Planckian and non-Planckian radiation. Furthermore, the modification of surface properties of materials under bombardment of intense low-energy ion beams can be studied.

Important areas for further research include study of stimulated Raman and Brillouin scattering in laser plasmas at higher intensities, nonlinear interaction of plasma instabilities, plasma opacities and equations of state (EOS), and plasma hydrodynamic stability and mixing. Research objectives within these areas include determining thresholds, saturation levels, and scaling relationships; normalizing high-temperature theoretical opacity models; obtaining high-pressure EOS data and standards for programmatically relevant materials; and characterizing phenomenology associated with plasma compression in excess of a factor of 100.

There also exists a related set of research areas that reflects the interdisciplinary nature of plasma science, when compared with the more formally recognized academic fields of study such as atomic physics, optical physics, condensed matter physics, and fluid mechanics. Consider the basic area of radiation-plasma interaction. ICF programs have used laser light to compress and heat thermonuclear fuel, magnetic fusion energy programs have considered electromagnetic radiation as a supplemental plasma heating source, and programs have been conducted to modify the ionosphere with high-power, high-frequency radiation. All exhibit basic common physical phenomena (such as instabilities, nonlinearity, turbulence, particle acceleration, and heating) and effects of spatial inhomogeneity (such as mode conversion, modification of instabilities, and wave propagation). Phase conjugation and wave mixing in plasmas acting as the nonlinear medium are examples of nonlinear optics phenomena whose analogues in optical media have led to previous technological applications.

The recent development of ultrashort-pulse lasers provides an opportunity to address basic science questions that cross over the descriptive boundaries of ICF plasma physics, atomic physics and condensed matter physics. For example, potential studies include behavior of atoms, ions, and molecules in the strong electromagnetic fields of the ultra-short-pulse laser; the properties of hot, solid-state plasmas; and radiation-matter interactions in all density-temperature regimes.

Early in research on inertial confinement fusion, laser-plasma experiments revealed many interesting plasma physics phenomena, including several types of parametric instabilities, superthermal particle acceleration, and spontaneous magnetic field generation. Ultrafast-pulse laser technology has now progressed to the point where extremely nonlinear phenomena are accessible. Incident electric fields as high as 10 kV/Å transform bound electrons instantly into relativistic free electrons. These so-called tabletop terawatt lasers are university-scale facilities.

Numerous new phenomena will be available for study. These include relativistic self-focusing, relativistic penetration into overdense plasmas, coupling that leads to energetic electrons and ions ($E > 1$ MeV), severe ponderomotive pressure modification of the plasma hydrodynamics, and coherent radiation generation. Large-amplitude plasmons can be created by several different processes (beat wave, wake field, etc.), and their evolution from coherent to chaotic structures can now be studied. Electron heating by these large-amplitude, high-phase-velocity waves can be examined, as well as the subsequent wave-particle interactions.

The short-pulse aspect (<1 ps) also enables studies of high-energy-density plasmas created through rapid energy deposition (~10^{30} W/g may be possible). The pulse duration may also be matched to characteristic time scales of the plasma, such as the period of an electron plasma oscillation for electron densities exceeding 10^{17} cm^{-3}. The fundamental properties of electromagnetic wave interaction change qualitatively, and numerous processes can lead to the acceleration of particles to high energy.

High-irradiance, short-pulse lasers may be able to generate extremely large magnetic fields (B >100 MG) in plasmas. Diagnostic techniques, such as subpicosecond Faraday rotation, hold the promise of measuring the evolution of such fields. Plasma resistivity can be controlled by the judicious choice of experimental parameters (including preformed plasmas), and fields initially containing more than 10% of the laser energy may be possible. Significant plasma flows can be created with high-brightness lasers. These plasma flows, which can have velocities approaching 10^9 cm/s, can be directed across externally applied magnetic fields or used to form counterstreaming plasmas. Numerous plasma phenomena can be studied.

The ultrafast regimes are particularly amenable to computer simulation and are an ideal test bed for improving the understanding of strongly driven plasmas. Research in this area could lead to new compact sources of tunable radiation and to a new class of compact particle accelerators. New diagnostic tools could be made available for material, biological, and electronic applications. Finally, improved understanding of strongly driven plasmas will benefit fusion and astrophysical research, as well as advance nonlinear science.

The equation of state of plasmas formed over a wide range of thermodynamic pathways can be studied at pressures exceeding 1 Gbar. High-irradiance, short-pulse lasers and high-intensity lasers can be utilized in these studies. The creation of strongly coupled plasmas and the near-isentropic compression of plasmas by lasers to densities exceeding 10^{25} cm^{-3} at temperatures less than the Fermi energy are now possible. The properties of such plasmas, of interest to inertial confinement fusion, astrophysics, and condensed matter physics, can now be studied in the laboratory. Experiments have been conducted demonstrating ultrahigh-pressure shocks exceeding 700 Mbar. Pressures readily in excess of 1 Gbar are now possible, enabling the equation of state and other aspects of condensed matter physics to be studied in regimes previously unattainable in the laboratory.

Finally, high-energy-density, non-LTE plasmas play a major role in several multidisciplinary endeavors, including optimized ICF target and driver design, x-ray plasma diagnostic spectroscopy, and x-ray lasers. Energy balance, hydrodynamic behavior, and radiation transfer are all affected by the detailed atomic states and kinetics of the plasma. Experiments would further normalize the computational ability to model and simulate plasma behavior and would further improve the spectroscopic ability to measure and characterize the plasma state.

CONCLUSIONS AND RECOMMENDATIONS

Certain trends are already evident in light of the evolving national priorities triggered by the commonly acknowledged end of the Cold War. First, a healthy process of consolidation and collaboration is evident. Increasingly, joint efforts addressing design, experiment, analysis, and facility issues collectively involve the Livermore, Sandia, Los Alamos, Naval Research, and University of Rochester laboratories. Technology collaboration and transfer arrangements with the industrial sector and a renewed emphasis on the quality and commitment to education are being discussed. Collaborative teaming and partnering among and within government, industrial, and academic organizations are being nurtured all the while in a fiscal environment focused on national deficit and debt reduction. It is in this context that recommendations regarding the funding and implementation of basic plasma science research impacting ICF are made.

A segment of the ICF community holds the view that the health of the field is adversely affected by the priority allocation of resources to facilities and operating costs, at the expense of support for basic plasma research aimed at a fundamental understanding of relevant phenomena. This resource allocation is reflective of an ICF program that relies primarily on computer simulation and full-scale experimental results. It also is suggested that the historical association between the ICF and nuclear weapon programs of the Department of Energy (DOE) has limited the size, nature, and degree of involvement of the basic plasma research community doing related work.

Past classification and facility access policies compounded this problem. However, recent DOE plans to declassify large portions of the ICF program provide a major opportunity to involve the basic plasma research community.

Given budgetary constraints, capital-intensive full-scale experimentation can constrain support for more fundamental theoretical and experimental scaling research and modeling. Although full-scale experimentation is essential, the inclusion of a basic plasma science research component within the fusion energy program can lead to a more timely achievement of the basic goals.

The ability to conduct basic plasma science research in ICF, as described above, depends critically on the accessibility of facilities and the availability of equipment, independent of the organizations and personnel involved. A dual approach is suggested. The previous policy of developing facilities for full-scale experimentation has put in place large numbers of components, subsystems, and equipment. The reconfiguration and recommissioning of smaller-scale research facilities should be considered to make effective use of existing equipment and capabilities.

Consideration should be given to providing the opportunity for a broader representation of participating organizations. Interested universities, small businesses, and corporate America could participate competitively, while at the same time offering cost-sharing opportunities. Existing federal programs, such as the National Laser User Facility (NLUF), the Small Business Innovation Research (SBIR) program, and Cooperative Research and Development Agreements (CRADAs) between industry and the national laboratories, could be helpful in this effort.

Consideration should be given to allocation of funding within the inertial confinement fusion program to support more related basic research and use of major ICF facilities as national user facilities. Where appropriate, ICF facility use should be encouraged in support of nonfusion programs. If no additional funding is available, basic plasma science research judged to be the most important could be funded from large facility accounts.

4

❖

Magnetic Confinement Fusion

INTRODUCTION

Plasma science has played a major role in magnetic fusion research from its inception and, in many ways, the quest for controlled fusion has been crucial in the development of modern plasma science. In a fusion reactor, a mixture of deuterium and tritium is ionized and the resulting plasma, which is confined by a magnetic "bottle," is heated to temperatures of the order of a few hundred million degrees centigrade. As illustrated in Figure 3.1, the deuterium and tritium nuclei would fuse upon colliding, thereby forming helium nuclei and very energetic (~14-MeV) neutrons. These neutrons may be captured in a thermalizing blanket and the energy used for electric power generation. The needs of magnetic fusion research required a far better understanding of collective interactions in plasmas than existed in the 1950s and 1960s. After the initial series of experiments, of particular concern was the gross magnetohydrodynamic stability of magnetic confinement configurations, the anomalous transport of energy and particles, and the heating and fueling of confined plasmas to reactor-relevant temperatures and densities. Some of the fundamental properties of collective interactions can be probed in relatively simple plasma configurations, the kind of basic experimental plasma research discussed in Chapter 8. However, many collective phenomena can be observed only in hot and dense plasmas in complex magnetic field geometries. The investigation of such phenomena required the development of new diagnostics to probe the properties of hot and dense plasmas, giving birth to experimental plasma research in fusion grade plasmas. Progress in all of these research areas will be required for ultimate success in controlled magnetic fusion. In the following sections, the panel concentrates on the so-called tokamak confinement concept. A tokamak is a toroidal plasma chamber in which confinement is produced by an axial magnetic field and a toroidal plasma current, usually driven inductively by a transformer. The panel also considers important non-tokamak confinement geometries, however.

MAGNETOHYDRODYNAMICS AND STABILITY

Introduction and Background

To achieve the densities and temperatures required for a successful thermonuclear reactor, a plasma must be contained by magnetic forces (such a confinement geometry is sometimes called a "magnetic bottle") for a sufficiently long time to produce net thermonuclear power. In the attempts to achieve this confinement, stability has emerged as one of the most important problems. A plasma confined by a magnetic field is not in thermodynamic equilibrium and therefore is potentially able to break out of the confinement system by a large variety of instabilities.

Past Achievements

During the past 10 years, great progress has been made in understanding the equilibrium and macroscopic stability properties of the tokamak plasma. The majority of tokamaks today routinely produce equilibria that are much more complicated than those of the original circular-cross-section "doughnut" concept. When a tokamak plasma is deformed from its axisymmetric equilibrium state, macroscopic MHD instabilities usually set in. The most virulent of these are the "ideal" instabilities, which tap the free energy associated with the pressure gradient or the plasma current. The unstable modes grow rapidly and can result in sudden loss of the stored plasma energy. Theoretical predictions of the stability boundaries for MHD modes have been corroborated by experiments. A significant achievement is the validation of the dependence of the beta limit (β is the ratio of plasma kinetic pressure to magnetic pressure) on plasma current, minor radius, and magnetic field. Stable operation regimes have been developed based on a good understanding of the dependence of MHD stability on global plasma parameters. In recent years, MHD research has begun to focus on the next level of understanding, namely, the impact of internal profiles on stability. The so-called second stable regime, for example, is a consequence of the localized high pressure creating a favorable "magnetic well" that stabilizes pressure-driven instabilities. Experiments have already shown that the beta limit can be doubled by optimizing plasma profiles. Furthermore, recent tokamak results have indicated a possible correlation between stability and confinement, offering the promise of an operational regime of enhanced stability and improved confinement.

A second class of important but less virulent MHD instabilities is the resistive instabilities. They allow the magnetic field lines to open up and reconnect and may be accountable for the degradation of confinement.

A third class of MHD instabilities is that driven by energetic particles. Experiments have observed instabilities in beam-heated plasmas that resulted in the ejection of energetic beam particles. Other experiments have used energetic beam ions to simulate the alpha particles in igniting plasma.

Of particular concern are the effects associated with self-generated plasma currents ("bootstrap" currents) owing to high plasma pressures at high temperatures and finite density gradients. High bootstrap current fractions (>50%) have been self-consistently calculated and found to be stable at reasonable values of β_p in major fusion devices. Nevertheless, as the pressure is increased, the plasma may become unstable, and this is observed in some of today's experiments. Once we develop a better understanding of these processes, there are plans to improve the stability at high pressures and high bootstrap current fractions in future tokamak experiments.

Future Prospects

In recent experiments, so-called toroidal Alfvén eigenmodes (TAEs) were driven unstable with neutral beams and high-energy particles driven by radio-frequency power. These TAE instabilities are important since they may be driven unstable by the alpha particles produced in the deuterium-tritium (D-T) fusion reaction. Theoretical calculations of fast particle destabilization thresholds for TAE modes are in reasonable agreement with the experimental results. Upcoming tritium experiments at the Tokamak Fusion Test Reactor (TFTR) in the United States and the Joint European Torus (JET) in Britain will test these models for the first time in tokamaks that have significant densities of alpha particles.

Despite the stability problems as the limits of plasma pressure and currents are approached, significant progress has been achieved by building larger and "smarter" machines. Tokamaks' confinement and stability continue to improve, and it is important to continue to improve the tokamak concept for eventual use as an economical power-producing reactor. The DOE is proposing to build the Tokamak Physics Experiment (TPX), a new "steady-state" national tokamak research facility at the Princeton Plasma Physics Laboratory. This device is now being designed by a national team of physicists and engineers. The plan is to start operation at the beginning of the next decade. One of the TPX's key objectives will be to push the stability limits by controlling the toroidal current profile with current-drive methods (see below).

TOKAMAK TRANSPORT

Introduction and Background

Since plasmas of sufficiently high density and temperature must be formed to accomplish thermonuclear ignition, understanding the transport of energy and particles is the key to the design of a fusion reactor. Virtually all tokamak experiments worldwide have demonstrated a common consistent level of energy and particle confinement in the so-called L-mode (L standing for "low confinement") regime of operation. In addition to L-mode, a variety of regimes have been observed and studied that have improved confinement, typically in response to changes in the plasma boundary conditions. The most ubiquitous of these is the H-mode regime (H standing for "high confinement"), which differs from L-mode primarily by having a transport barrier at the plasma edge. In the H-mode, the energy confinement is typically a factor of two better than in the L-mode. More recently, even better confinement has been achieved in the so-called VH-mode ("very-high") regime, where confinement times up to four times longer than those of L-mode plasmas have been achieved. Characteristics of the H-mode and VH-mode are illustrated in Figure 4.1.

Past Achievements

During the past decade, tokamak plasma parameters comparable with those estimated to be required for fusion reactors have been achieved (Figure 4.2), including maximum central ion temperatures of $T_i \sim 37$ keV (4×10^8 °C), confinement times of $\tau \sim 1$ s, and central densities $n \sim 3 \times 10^{20}$ m^{-3}. These parameters were not all achieved simultaneously, but simultaneous measurement has been achieved of a fusion triple product $n\tau T_i \sim 1.1 \times 10^{21}$ keV s m^{-3}. Current experiments operate with the same dimensionless parameter values as those expected in reactors, with the exception that reactors will require two to three times greater confinement times and electron temperatures than those obtained in today's machines.

Tokamak experiments in the 1980s demonstrated that the transport of thermal plasma energy and momentum across the confining magnetic field is ~100 times faster than predicted by theory. This theory considers only the effects of Coulomb collisions and essentially is an extension of gas-dynamic models to magnetized plasma transport. These observations have shifted the emphasis in tokamak cross-field transport theory from collisional models to ones that include the effects of plasma turbulence.

Dimensionless scaling experiments have begun measuring the scaling of tokamak cross-field thermal transport with respect to the theoretically important dimensionless parameters, in a manner analogous to windtunnel experiments. It has been found that the scaling is unlike that predicted by most theories. In

PLATE 1 Three 82-mm-diameter silicon wafers are shown during plasma processing in a plasma reactor used for the transfer of fine-line features during electronic chip fabrication. The bright regions above the wafers are clouds of fine particles, illuminated by argon laser light. This "dust" is a critical source of defects in chip manufacture, and techniques are now being developed to eliminate it. (Courtesy of G. Selwyn, Los Alamos National Laboratory; work performed at IBM, Yorktown Heights, N.Y.)

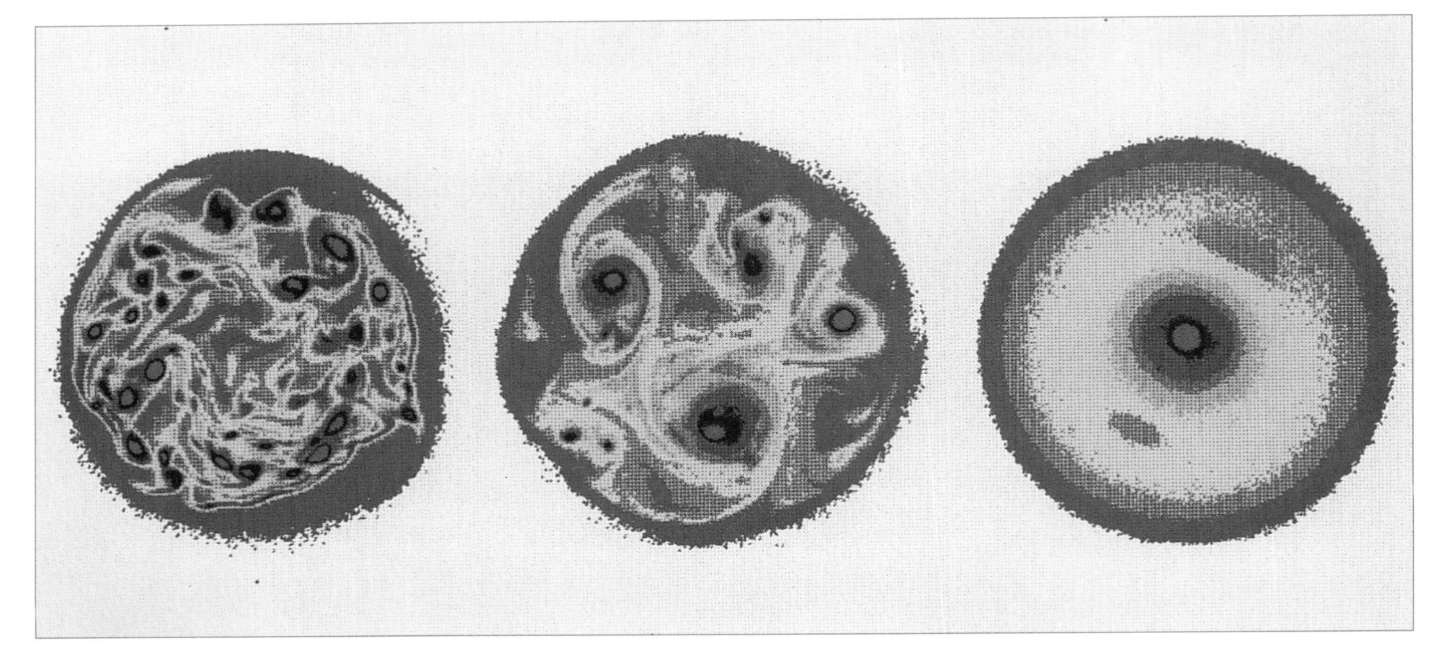

PLATE 2 Relaxation and self-organization of turbulence in a magnetized pure-electron plasma, showing the time evolution of fine-scale vorticity into one large vortex. These electron plasmas are nearly ideal, inviscid, two-dimensional fluids, in which electron density plays the role of vorticity. They permit studies of two-dimensional fluid dynamics not possible in conventional fluid systems. Shown is the evolution of electron density, as time proceeds, from left to right; violet corresponds to high electron density (i.e., vorticity) and red to low density. (Courtesy of C.F. Driscoll, University of California, San Diego.)

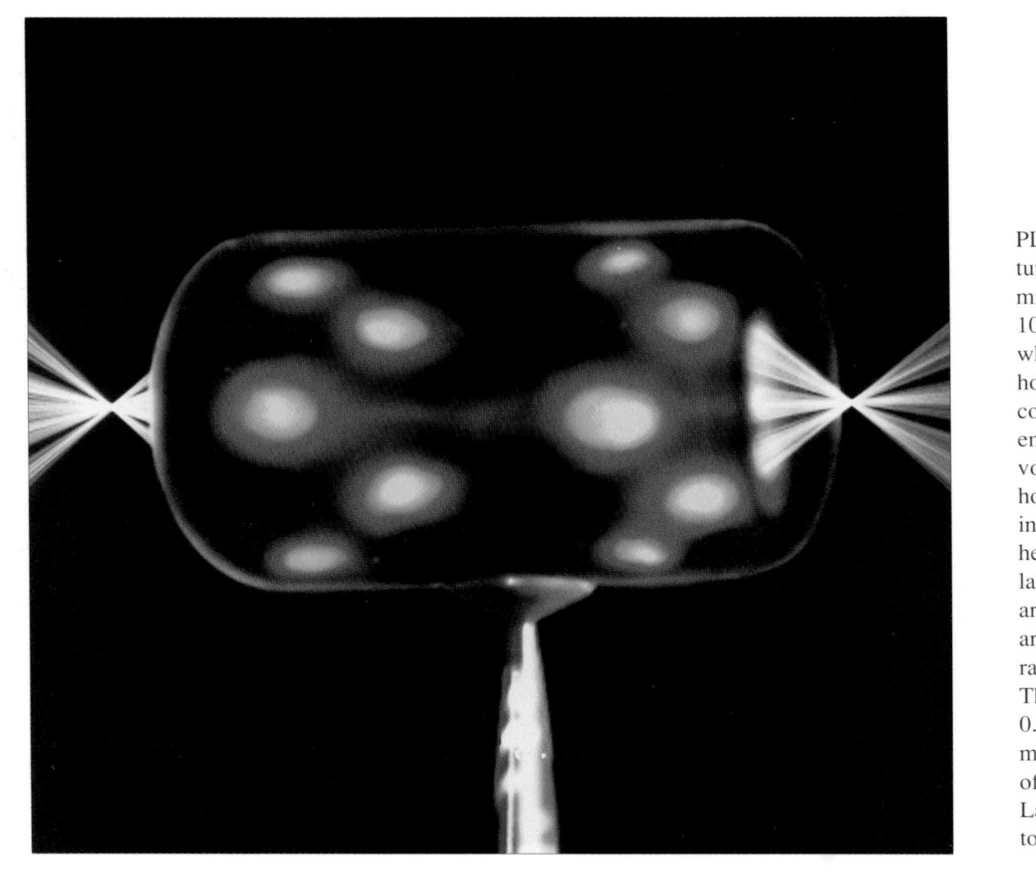

PLATE 3 An ICF hohlraum (a thin tungsten cylinder 2.8 mm long and 1.6 mm in diameter) is illuminated by the 10 beams (in blue) of the Nova laser, which enter through small holes in the hohlraum. In this approach to inertial confinement fusion, intense x-rays with energies of several hundred electron volts are generated and trapped in the hohlraum when the laser beams heat its interior walls. These x-rays uniformly heat the ICF target. In this figure, the laser beams, hohlraum, and stalk are the artist's conception. The orange spots are actual experimentally measured x-ray emission from the hohlraum walls. The laser beams have a wavelength of 0.35 μm. The spots are each approximately 0.7 mm by 0.3 mm. (Courtesy of F. Ze, S. Wilks, and S. Dougherty, Lawrence Livermore National Laboratory.)

PLATE 4 X-ray laser interferogram of a mylar target irradiated with 1.5×10^{14} W/cm^2 by one beam of the Nova laser. To produce this image, an x-ray laser operating at a wavelength of 155 Å, with a pulse duration of 150 ps, was combined with a Mach-Zehnder interferometer consisting of multilayer mirrors and multilayer-coated silicon-nitride beamsplitters. Compared to conventional optical interferometry, the short wavelength of the x-ray laser makes it possible to probe much larger and higher-density plasmas with micron resolution. In this image, one fringe shift corresponds roughly to an electron density of 1.5×10^{20} cm^{-3} for a plasma length of 1 mm. The bright region is self-emission from the laser-heated plasma. (Courtesy of Luiz Da Silva, Lawrence Livermore National Laboratory.)

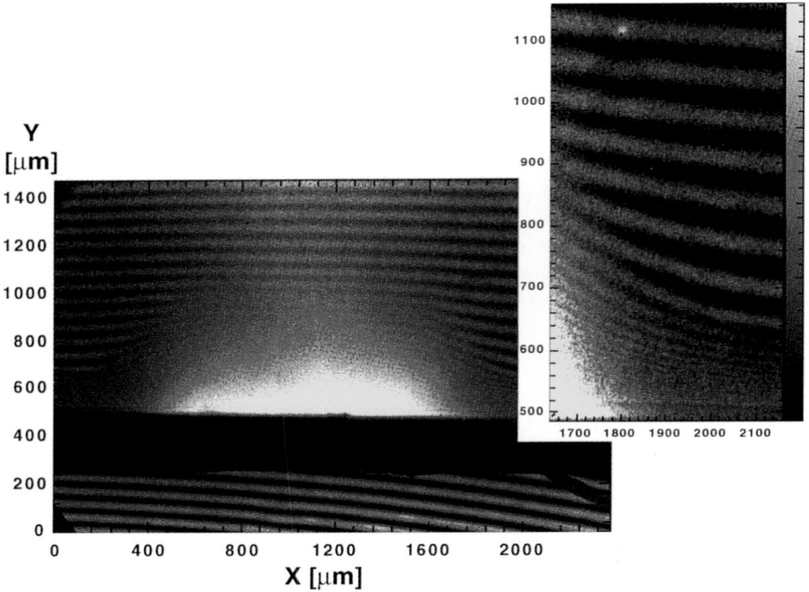

PLATE 5 (facing page below) Computer model of Earth's plasma environment, obtained by solving the single-fluid magnetohydrodynamic equations on a supercomputer. This computation illustrates the breadth of phenomena that can be simulated using contemporary hardware and software. The terrestrial magnetic field lines are compressed and confined on the sunward side by the solar wind, which flows in from the left (i.e., the day side) and is drawn out into a long magnetotail at night. Note the presence both of magnetic field lines connected to Earth and of lines that are completely disconnected, and the juxtaposition of the two at points, both day and night, where magnetic reconnection presumably occurs. Coloring denotes relative plasma pressure, which is high (red) in the interface region with the solar wind (the magnetosheath) and in the near-Earth nightside (the plasma sheet) and is very small (blue) in the distant tail lobes. The size of Earth has been artificially magnified so that the distribution of aurorae, which are proportional to the incident plasma flux, is readily discernible. (Courtesy of J. Raeder, University of California, Los Angeles.)

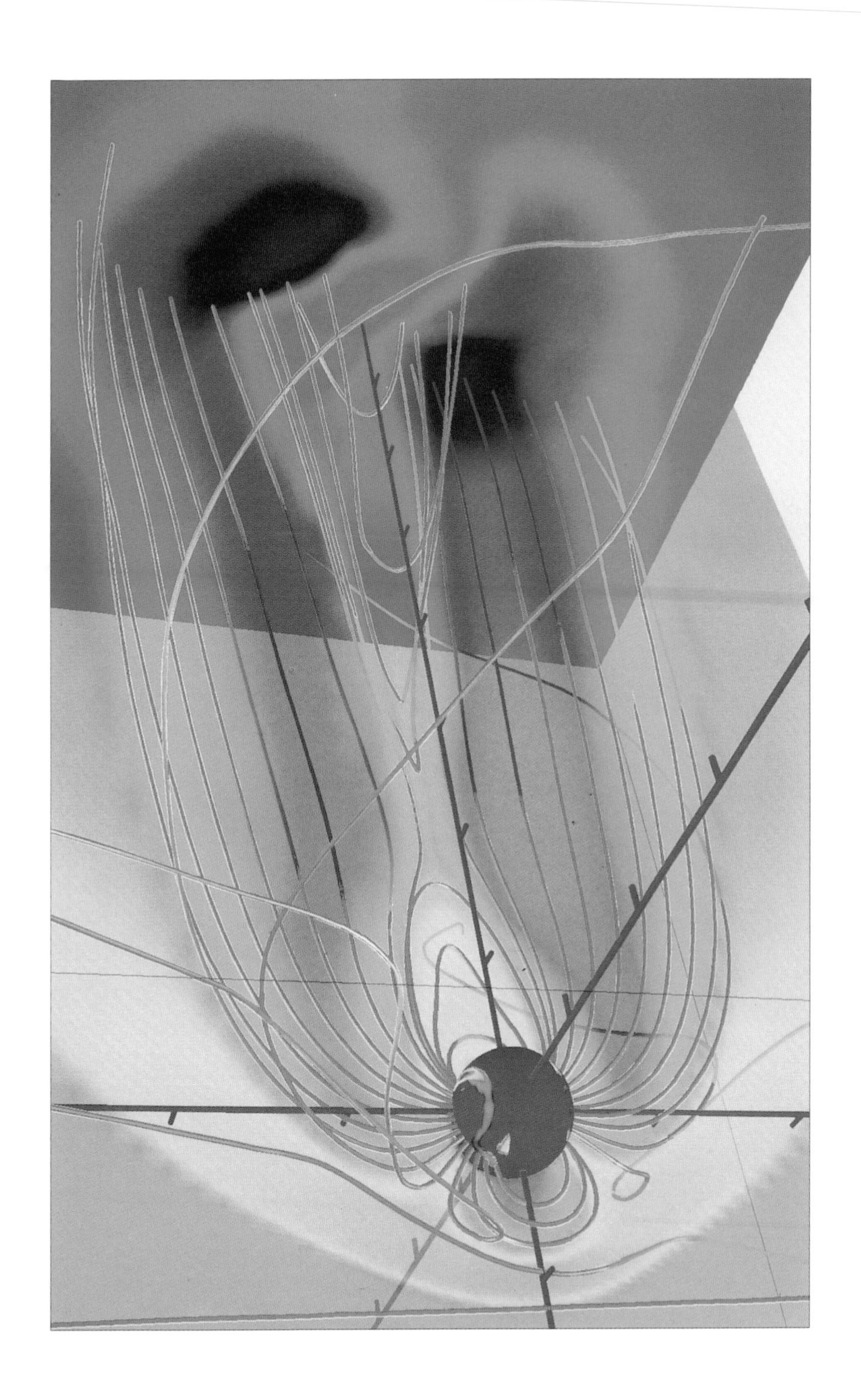

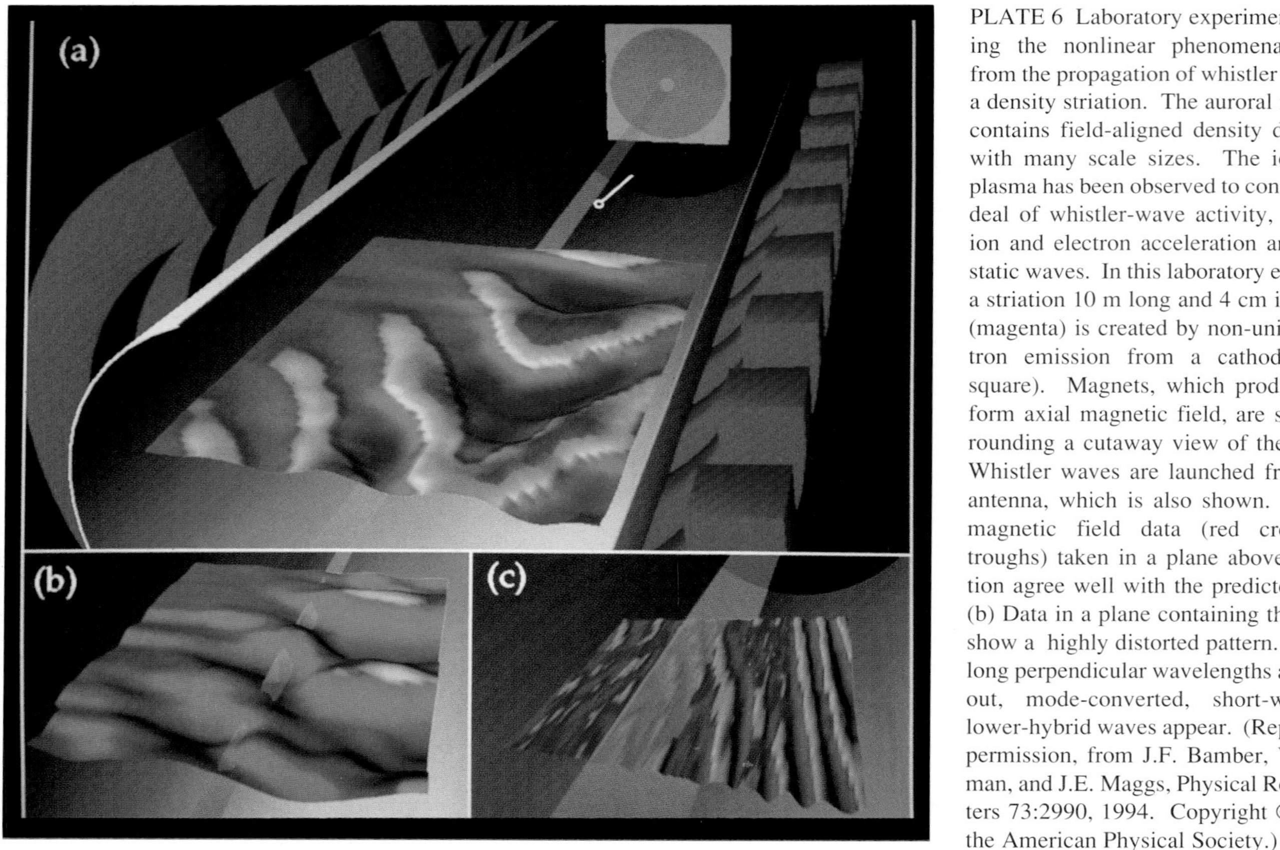

PLATE 6 Laboratory experiment illustrating the nonlinear phenomena resulting from the propagation of whistler waves into a density striation. The auroral ionosphere contains field-aligned density depressions with many scale sizes. The ionospheric plasma has been observed to contain a great deal of whistler-wave activity, as well as ion and electron acceleration and electrostatic waves. In this laboratory experiment, a striation 10 m long and 4 cm in diameter (magenta) is created by non-uniform electron emission from a cathode (orange square). Magnets, which produce a uniform axial magnetic field, are shown surrounding a cutaway view of the chamber. Whistler waves are launched from a loop antenna, which is also shown. (a) Wave magnetic field data (red crests, blue troughs) taken in a plane above the striation agree well with the predicted pattern. (b) Data in a plane containing the striation show a highly distorted pattern. (c) When long perpendicular wavelengths are filtered out, mode-converted, short-wavelength lower-hybrid waves appear. (Reprinted, by permission, from J.F. Bamber, W. Gekelman, and J.E. Maggs, Physical Review Letters 73:2990, 1994. Copyright © 1994 by the American Physical Society.)

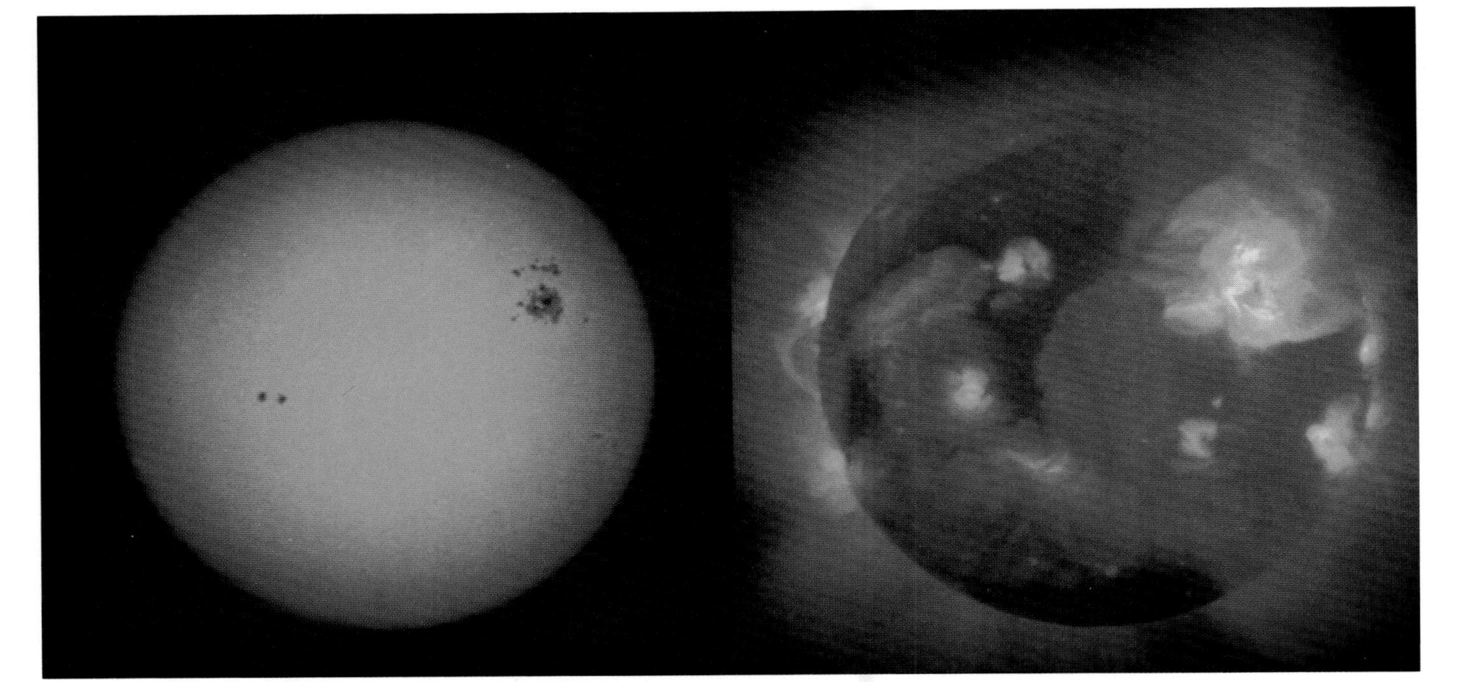

PLATE 7 Images of the Sun in white light (left) and in soft x-rays (right) taken by the Yohkoh spacecraft in 1991, illustrating the importance of magnetic fields in solar phenomena. The white light shows that except for active regions, which are distinguished by sunspots, the Sun is essentially featureless at the 5000 K photosphere. However, as evidenced in the x-ray image, the million-degree corona exhibits a great amount of structure due to the influence of the magnetic field. The corona is most active over sunspot regions. At the poles, where the magnetic field is weak and open, x-ray emission is minimal. (Courtesy of Lockheed Palo Alto Research Laboratory, NASA, and the Institute of Space and Astronautical Science of Japan.)

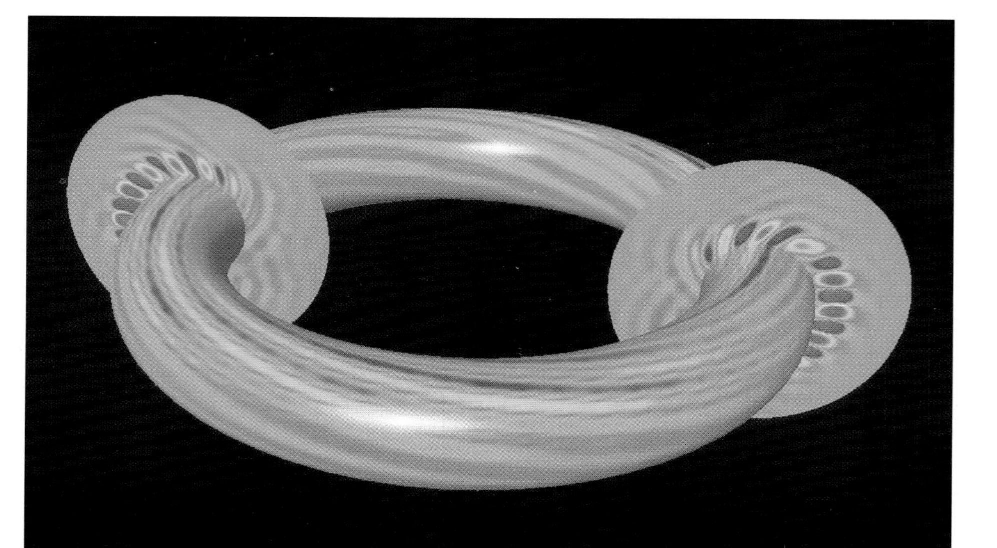

PLATE 8 Computer simulation of plasma turbulence in a tokamak plasma, using a gyrokinetic approximation and 1 million to 8 million particles. Radial and poloidal cross sections of the electrostatic potential are shown during the linear amplification phase. This picture highlights the radial elongation of the mode structure during saturation and the transition to turbulence. The colors represent the value of the potential, varying from positive to negative: magenta, violet, blue, green, yellow, red. (Reprinted, by permission, from S.E. Parker, W.W. Lee, and R.A. Santoro, Physical Review Letters 71:2042, 1993. Copyright © 1993 by the American Physical Society.)

particular, the experiments imply that the size of the turbulent eddies is not always controlled by the electron or ion gyroradius size, but presumably is set by macroscopic scale lengths. New diagnostic techniques used in recent measurements of the turbulent fluctuations within tokamak plasmas have found that the spectrum and implied transport are dominated by moderately long-wavelength modes. Measured changes in the plasma transport are well correlated with changes in the fluctuation amplitude, implicating them as the cause of the transport. In contrast, the measured transport parallel to the magnetic field is in good agreement with the predictions of "neoclassical" theory.

One of the major scientific achievements of tokamak transport and turbulence studies has been the development of a model explaining the formation of the transport barrier in H-mode. This model is based on stabilization of turbulence by sheared $\boldsymbol{E} \times \boldsymbol{B}$ flow (here $\boldsymbol{E}$ is the radial electric field observed in the vicinity of the tokamak edge, and $\boldsymbol{B}$ is the toroidal magnetic field). The measured levels of $\boldsymbol{E} \times \boldsymbol{B}$ flow shear are well above those theoretically required for such stabilization. Furthermore, the increased $\boldsymbol{E} \times \boldsymbol{B}$ flow shear is correlated with the reduction of density fluctuations, cross-field energy, and particle transport both spatially and temporally. These results provide some of the best evidence to date of the close connection between fluctuations and transport. Similar $\boldsymbol{E} \times \boldsymbol{B}$ shear stabilization effects may also take place in the core of tokamak plasmas (e.g., the confinement improvement from H-mode to VH-mode).

The concept of electric field flow shear stabilization of turbulence may be one of the most fundamental contributions of tokamak physics to general fluid dynamics. Although flow shear stabilization can take place in ordinary fluids, a sheared velocity field is usually a source of free energy; hence it usually drives instabilities rather than stabilizing them. Only in a plasma can magnetic shear prevent instabilities driven by velocity shear (e.g., Kelvin-Helmholtz instabilities) so that flow shear can then affect the other instabilities.

Another recent result is the importance of the plasma current density profile in controlling confinement. Experiments modifying the current profile transiently, by inductively driving a skin current, changing the plasma shape, or using external current drive, have shown that, with peaked current profiles, the cross-field transport can be substantially reduced. Although this effect is not understood theoretically, measurements of the local ion thermal transport indicate that it may be reduced by increasing magnetic shear.

Future Prospects

The eventual goal of the magnetic fusion program is the realization of a commercial reactor to generate electricity. Present limitations on confinement and total pressure in the plasma force reactor designers in the direction of multigigawatt units. Understanding and reducing turbulent transport in toka-

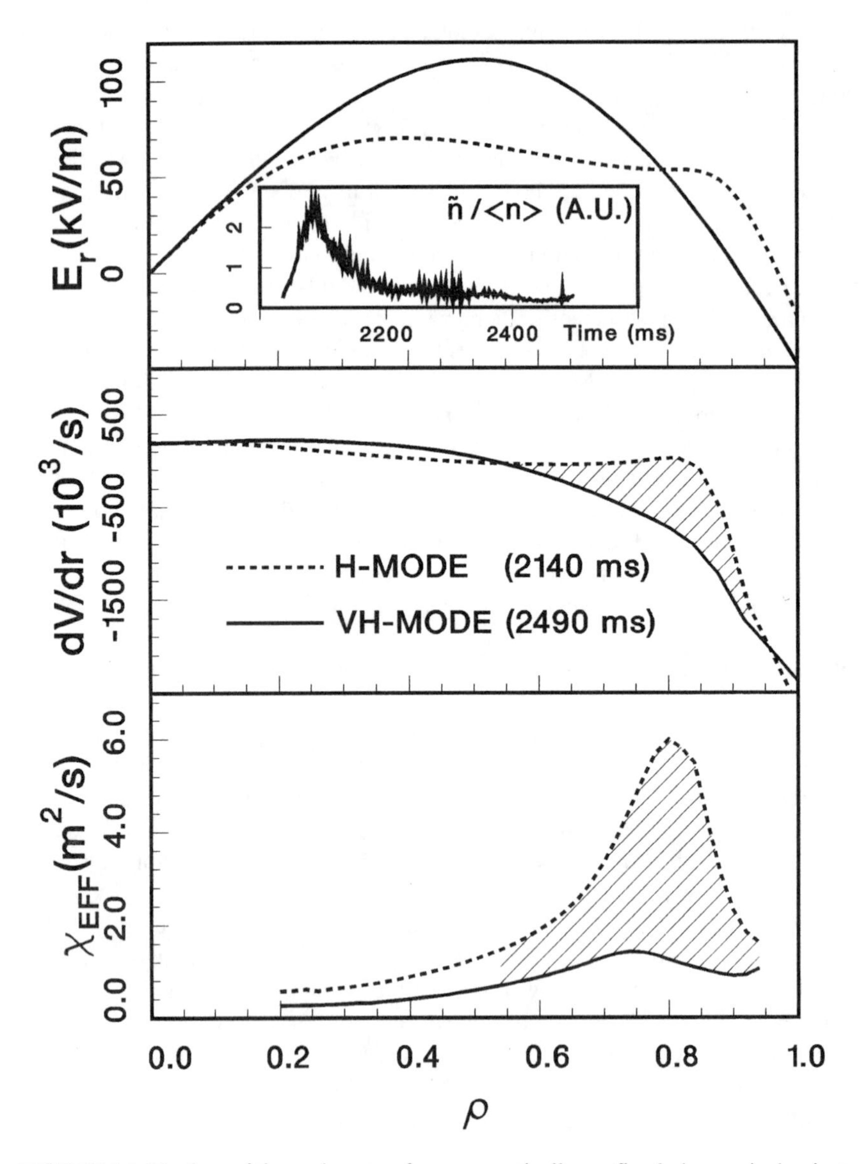

FIGURE 4.1 The loss of thermal energy from magnetically confined plasmas is dominated by turbulent transport. The high-confinement regime (H-mode) in tokamaks is characterized by reduced turbulence at the plasma edge, which inhibits this outward transport of energy. The existence of this regime is believed to be caused by a shear in the drift velocity, associated with a radial electric field. Recently, an even higher confinement regime (the VH-mode) has been discovered, in which the region of high shear and low turbulence extends deeper into the plasma. Shown in this figure are, from top to bottom, the electric field, the velocity shear, and the thermal diffusivity as functions of the nor-

maks could have a major payoff in reduced-size commercial reactors in the next century.

A detailed theoretical understanding of tokamak turbulent transport from first principles continues to elude us. However, continued development and exploitation of novel diagnostics and experiments, coupled with the new nonlinear numerical simulations, should allow identification of the dominant turbulence drive and damping mechanisms during the next decade. Theoretical understanding of the turbulent transport mechanism should allow the development of new techniques for controlling transport to improve tokamak reactors.

Present experiments have shown that control of the plasma current and electric field (flow shear) profiles can have significant effects on turbulence and confinement. Similarly, optimizing the current profile, pressure profile, and plasma shape are important for increasing the beta limit. However, in many cases present experiments in current profile control have been done with techniques that are inherently transient. They demonstrate the principle, but they are not necessarily sufficient to prove that such profiles can be maintained in steady-state or in long-pulse reactors. The challenge for future experiments will be to demonstrate active control of the plasma current, electric field, and pressure profiles by techniques that are economical and applicable to long-pulse or, preferably, steady-state devices.

EDGE AND DIVERTOR PHYSICS

Introduction and Background

Divertors are magnetically separated regions at the boundaries of magnetic confinement fusion devices. Their originally envisioned functions (Spitzer, 1957) were to exhaust heating power and helium "ash" from fusion reactions and to protect the reacting core plasma from impurities. A separatrix is a surface that separates the flux region of the core plasma from the "burial chamber," the region remote from the core plasma where magnetic field lines intercept material surfaces. The plasma that diffuses across the separatrix from the hot core is "scraped off" on those material surfaces, thus giving rise to the name scrapeoff layer. The edge plasma is located just inside the magnetic separatrix and the

---

malized minor radius, ρ, for the H- and VH-modes. The inset at the top of the figure shows the time evolution of the normalized density fluctuations, as measured by microwave scattering. (Reprinted, by permission, from K.H. Burrell, T.H. Osborne, R.J. Groebner, and C.L. Rettig, *Proceedings of the 20th European Physical Society Conference on Controlled Fusion and Plasma Physics*, European Physical Society, Geneva, 1993, vol. 17C, part 1, pp. 1-6. Copyright © 1993 by the European Physical Society.)

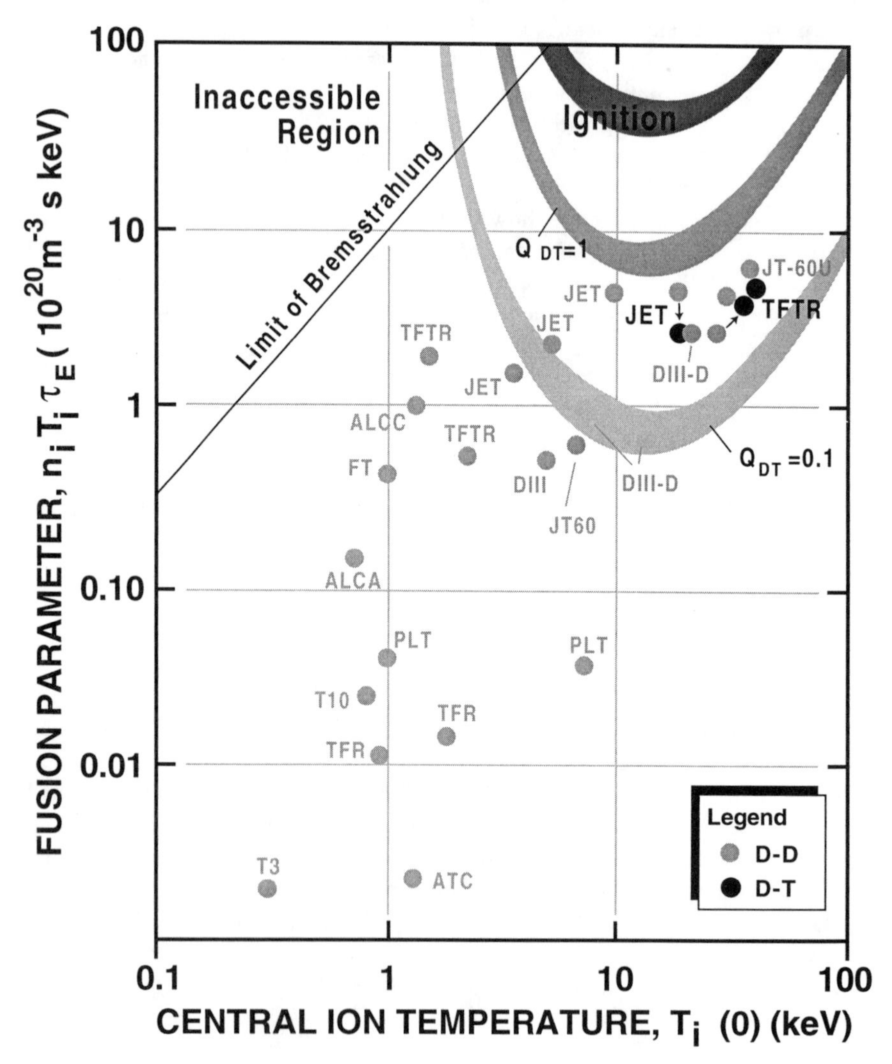

FIGURE 4.2 The approach to power plant conditions of magnetic confinement fusion experiments: The performance of various plasma devices is shown as a function of the "fusion parameter," $n_i T_i \tau_E$, and the central ion temperature, T_i. The boundaries are indicated for Q = 1 in a deuterium-tritium plasma (i.e., "scientific break-even," where fusion power out equals power in) and for ignition.

scrapeoff layer is just outside. The properties of the edge and scrapeoff plasma regions have been shown to exert a profound influence on the confinement and transport properties of the main plasma. This arises through modifications to the boundary conditions on charged and neutral particle and energy flows.

Recent Advances

Research on divertor physics was invigorated in the early 1980s by the discovery of an enhanced core plasma confinement mode (H-mode) for auxiliary-heated plasmas. A doubling of energy confinement was soon verified in many subsequent divertor experiments and, more recently, in limiter-bounded plasmas, as well as in stellarators (an alternate form of toroidal confinement device). New regimes of enhanced confinement were also discovered in ohmically heated tokamaks and in auxiliary-heated tokamaks with special vessel wall treatments. The common link between all these improved regimes of energy confinement was control over hydrogen recycling in the edge plasma.

To understand the transport of energy in the edge region, two-dimensional fluid modeling of the scrapeoff layer has yielded important predictions. A regime of high recycling divertor (HRD) was predicted in which the flux of particles onto the divertor plate greatly exceeded the flow of particles out across the separatrix. This greatly reduces the average energy of the ions hitting the divertor plate and, hence, the impurities generated there by sputtering. A flux enhancement factor of ~20 has been measured in several tokamaks, which confirms the model. In addition, the flux enhancement in the divertor acts as a flow against which impurities created at the divertor surfaces must fight to reach the plasma core region.

Though not directly within the category of plasma edge physics, activities on wall conditioning deserve special recognition. New methods to coat and condition the walls of tokamaks (e.g., carbon, boron, silicon, and lithium coatings and helium discharge conditioning) have resulted in the greatest improvements in core plasma phenomena. At present, this activity is more of an art than a science and is not well understood.

Future Research and Technical Opportunities

The requirements of the International Thermonuclear Experimental Reactor (ITER) program have given new momentum to edge and divertor physics research. The ITER activity, supported by the European Community, Japan, Russia, and the United States, is now in the engineering design phase. ITER will be the first large-scale device built for the purpose of demonstrating controlled thermonuclear ignition and burn as well as the engineering feasibility of a tokamak-based reactor. However, analysis by the ITER team has shown the inability of present conceptual designs of the divertor to handle the intense power loads with sufficient safety margin or lifetime.

Attention has returned to certain old, though still-untested, ideas and recent variations on how to widen the thin power-carrying scrapeoff layer. Ergodization of field lines is under scrutiny. Another scheme relies on the injection of cold gas into the scrapeoff layer to reduce, by charge exchange, the power carried to the divertor plate by plasma ions. Intense radiation, mainly by impurities entrained in the plasma flow toward the divertor plate, may be able to drain the power out of the electron channel. At the extreme, a completely successful embodiment of these approaches would result in volumetric plasma recombination and the replacement of a solid divertor plate by a gas target (an idea that originated in 1974).

The fluid codes presently used to model the scrapeoff layer do not represent a sufficiently accurate description of the physical processes. These must be improved by the direct incorporation of kinetic effects, drift motion, nonambipolar flows, and better atomic physics. With such improvements, these codes could better evaluate helium flows and impurity effects. Comparisons with helium exhaust experiments would prove enlightening. Methods to control helium flows by interactions with waves and MHD activity appear promising. Such activities are under consideration both for U.S. tokamaks and in collaborations on foreign tokamaks.

PLASMA HEATING AND NON-INDUCTIVE CURRENT DRIVE

Neutral-Beam Heating and Current Drive

Introduction and Background

Neutral-beam injection has been the principal method of heating the past several generations of tokamak plasmas and has also found utility in driving toroidal plasma currents. A neutral-beam injector consists of a high-current ion source, with multiaperture grids that electrostatically accelerate ions into a conductance-limiting duct, where a portion of the ions charge-exchange with gas to become fast neutrals. The residual ions are swept out of the beam by a deflection magnet, leaving the high-energy neutrals to pass through a duct across the tokamak's fringing magnetic fields into the plasma. Once inside the plasma, the beam neutrals are ionized, and these high-energy ions are captured by the magnetic field. As they circulate many times around the plasma, they collide with the plasma electrons and ions, transferring energy to them until the beam ions are thermalized. If the neutral beams are injected predominantly in one direction tangential to the plasma's major radius, they can drive net plasma current, reducing the need for inductive current drive after the plasma startup phase.

Past Achievements

Neutral-beam injection has several features that have made it attractive for implementing experiments on numerous tokamaks over the past two decades. Since the beam is trapped by the plasma and transfers energy to it through two-body interactions, the physics involved is relatively straightforward and calculable, allowing the power deposition profile to be predicted accurately. The experimental flexibility has enabled neutral-beam injection to provide most of the heating for transport and confinement studies on tokamaks since the 1970s. Over this period, the injected power levels have been increased from the 100-kW range to 33 MW. Recently, neutral beams have produced ion temperatures up to 44 keV in some of the world's major tokamaks. Beams, including tritium, were used in some cases to produce more than 10 MW of fusion power. (See Figure 4.3.) Beams played an essential role in the discovery of peaked density profile enhanced confinement regimes (called Supershots) and the first demonstration of "bootstrap current," a gradient-driven self-current, on a tokamak. Neutral beams also are important for driving plasmas into H-modes and VH-modes.

Future Prospects

The neutral beams used on tokamaks over the past 20 years have all been based on positive ion sources (with an electron added in the neutralizer to form the neutral beam). The practical neutralization efficiency that can be achieved with positive ions decreases very rapidly at deuterium beam energies of 120 keV and greater. Therefore, for future applications (MeV energies may be desirable in reactors), negative ion beams are more attractive. The achievable neutralization efficiency in an optimized-thickness gas cell is high for negative ion beams (58-60%) and is nearly independent of energy in the hundreds of keV to many MeV range. The roles for neutral beams in the future will include reliable plasma heating and central plasma current drive. Technical opportunities abound for improving the current density, brightness, and gas efficiency of negative ion sources, and for perfecting photodetachment neutralizers and plasma neutralizers that could permit still higher neutralization efficiencies.

Radio-Frequency Heating and Current Drive

Introduction and Background

An alternative way to heat plasmas to high temperatures is by means of radio-frequency waves. Radio-frequency heating spans a very large range of frequencies, from a few megahertz (MHz or 10^6 Hz) to a few hundred gigahertz (100 GHz or 10^{11} Hz). (One hertz designates one cycle per second oscillation frequency.) The low-frequency end corresponds to the regime of Alfvén waves,

FIGURE 4.3 Photograph of the Tokamak Fusion Test Reactor (TFTR), located at the Princeton Plasma Physics Laboratory. The major radius of the donut-shaped plasma is 2.5 m. Typical plasma currents are 2.5 MA at toroidal magnetic fields of 52 kG. Powered by intense neutral beams and with a deuterium-tritium fuel mixture, TFTR has achieved record ion temperatures of 44 keV and fusion powers of 10.7 MW in second-long pulses. (Courtesy of Princeton Plasma Physics Laboratory.)

while the high-frequency regime corresponds to electron cyclotron waves, which are resonant with electrons gyrating at their gyro (cyclotron) frequency or its harmonics. Other frequencies of interest include the ion-gyro frequency or its harmonics (30-200 MHz) and the ion plasma frequency (more accurately the so-called lower-hybrid frequency) at 1-4 GHz. The basic premise of rf heating is that an antenna installed in the vicinity of the vessel wall radiates electromagnetic waves that deliver rf power from a transmitter to the high-temperature plasma core where the power is absorbed by wave-particle resonances. For example, Alfvén waves and lower-hybrid waves may transfer their energy and momentum to electron motion parallel to the magnetic field by the process of "Landau damping" (named after the famous Russian theoretical physicist who first predicted this kind of resonant wave-particle interaction nearly five decades ago). The accelerated resonant particles eventually dissipate their energy in the background plasma by collisions, thereby heating the bulk plasma particles. If, in addition, the waves travel in a preferred toroidal direction (which can be arranged by proper phasing of the antenna elements), net momentum is transferred to charged particles (usually electrons), thereby generating a net toroidal plasma current. Since rf generators can be operated continuously (cw, or continuous wave operation) at the megawatt power level, a steady-state tokamak operation may become feasible.

In spite of its complexity, the physics of waves in plasmas presents one of the best-understood and most scientifically established disciplines in plasma science. Many aspects of this subject have been verified in fundamental "basic physics" experiments, while some aspects are still under study in the complex field geometry of toroidal configurations. However, the nonlinear aspects of wave propagation still are not well understood.

Past Achievements

Over the past decade, the scientific discipline of radio-frequency heating and current drive in plasmas has seen rapid growth both in the United States and abroad. As in the case of neutral-beam heating, most rf experiments on fusion plasmas in the 1970s consisted of sources amounting to only a few hundred kilowatts, increasing to the ~1-MW level in the early 1980s and culminating recently at the 22-MW injected power level in the ion-cyclotron (ICRF) frequency range. Impressive heating results have been obtained recently in large tokamak devices, where central electron temperatures up to 13 keV have been achieved. In agreement with theoretical projections, directly accelerated "minority ion" species with up to MeV energies have been observed. Detailed energy analysis of these energetic ions shows excellent agreement with "quasilinear" wave-particle interaction theories and large simulation codes, one of the triumphs of modern plasma theory.

Perhaps even more striking results have been obtained in the area of noninductive current drive by rf waves. In this case, the waves not only heat the plasma (i.e., transfer wave energy to particles) but also transfer net momentum in the toroidal direction. In the past decade, current drive by rf waves has been verified in nearly all frequency regimes. Recently, in Japan, currents at the 3.5-MA level have been driven by multimegawatt lower-hybrid waves. Since these currents are often driven by electrons with energies of 100s of keV, it has been possible to study the transport of these energetic electrons by x-ray imaging techniques. Another scientific spinoff of these experiments is a better understanding of stochastic acceleration of charged particles in electromagnetic wave fields. This may have important application to astrophysical and space plasmas.

Future Prospects

Radio-frequency heating and current drive will likely be used in all future toroidal plasma devices. While ICRF power is eminently suitable for bulk plasma heating, even under reactor-like conditions, electron cyclotron resonance heating may be used either for bulk plasma heating or for special-purpose localized heating (temperature profile control). In principle, localized heating offers the possibility of controlling the pressure profile and thereby improving MHD stability. Lower hybrid waves have been most successful at driving toroidal plasma currents (lower-hybrid current drive, or LHCD). In future experiments, LHCD will be used mainly for driving off-axis plasma currents (current profile control), while the central currents may be driven with neutral beams or with fast magnetosonic waves in the ICRF regime. For purposes of disruption control, highly localized edge current drive with electron cyclotron waves is contemplated.

In the future, rf wave theory will concentrate on the nonlinear regime. In many cases this leads to the study of strongly nonlinear regimes in plasmas, including turbulence, chaos, and stochastic particle acceleration. An understanding of these phenomena will have a large impact on our understanding of similar phenomena in astrophysical and space plasmas, including solar physics.

Radio-frequency heating is a strong technology driver. Higher-power radio-frequency sources are under development in nearly all frequency regimes. In the ICRF regime, new tetrodes have been developed by industry with cw power levels up to 3 MW; future directions include the possible development of 5-MW tube capability. In the lower-hybrid regimes, cw tubes (klystrons) up to 0.5 MW have been developed at 2.45-3.7 GHz, and for future applications, 1 MW tubes are a good prospect for development. Finally, in the electron-cyclotron resonance heating (ECRH) regime, ~1-MW pulsed tubes (gyrotrons) have been developed at frequencies in the 100-GHz range, and future development work promises cw tubes at the 1-MW level at frequencies up to 150 GHz.

DIAGNOSTIC DEVELOPMENT

Introduction and Background

The need to measure detailed plasma parameters in fusion-grade plasma environments has led to many creative applications of plasma science. In turn, numerous advances in plasma science have been inspired directly by the need to understand plasma properties with ever-increasing precision. Important examples from the last decade of research are indicated below. In addition, a summary of future directions is presented.

Past Achievements

Density and electron temperature profiles are routinely measured by laser Thomson scattering and laser interferometry. Electron temperature profiles in hot plasmas are also measured by electron cyclotron emission (ECE). Ion temperatures are often measured by spectroscopic techniques. Among more novel diagnostics, one of the most important and, therefore, most intensely investigated areas is that of incoherent fluctuations and their relationship to energy and particle transport. The fluctuating quantities of interest include density, temperature, and plasma potential. To relate these to transport, it is necessary to measure frequency and wavenumber spectra, along with relative phase, as functions of spatial location. Many techniques have been developed to attack these problems. Beam emission spectroscopy (BES) allows for the measurement of electron density fluctuations by looking spectroscopically at line radiation that results from plasma excitation of high-energy neutral-beam atoms injected into the plasma. BES has led to the discovery of large-scale structures that are now the subject of intense theoretical investigations. Other techniques that have been invented to measure density fluctuations include reflectometry (backscattering from the critical layer), laser scattering, and phase-contrast imaging. Utilizing heavy-ion beam probes of very high energy (~1 MeV), measurements of both density and potential fluctuations have been carried out. Probes have long been used in low-temperature plasmas to measure both density and potential fluctuations, and these are used routinely in the scrape-off region of tokamak plasmas. For the first time, fast scanning probes have allowed access to hotter regions of plasma, inside the last closed flux surface. The ability to measure density and potential fluctuations simultaneously has allowed the first direct measurements of fluctuation-induced energy and particle cross-field transport.

The desire to know the detailed structure of the magnetic field in the hot confinement region of tokamaks has spawned several creative new diagnostic techniques. These include the application of Faraday rotation, Zeeman polarimetry using neutral beams and pellets, and the imaging of the ion clouds that result from pellet ablation. The approach that probably has the most potential for highly precise internal field measurements with good spatial resolution involves an application of the motional Stark effect (MSE), also using an energetic neutral beam. This has led to detailed q profile (safety factor) measurements and new insights into MHD phenomena.

The ability to measure detailed *profiles* of plasma parameters also has matured significantly over the last 10 years. By combining measurements from multiple arrays of soft x-ray sensitive diode detectors with new tomographic inversion algorithms, a wealth of new physics information on the structure and evolutions of fusion plasmas has been gleaned. Charge exchange recombination spectroscopy (CXRS), whereby excited states of hydrogen-like ions of low-Z impurities, such as carbon and oxygen, are populated by charge transfer from atomic hydrogen beam atoms, has enabled detailed local measurements of ion temperature profiles. This is crucial to our attempts to understand the mechanisms responsible for cross-field energy transport. Perhaps even more important, this approach has allowed for the measurement of plasma rotation and, particularly near the edge of plasma, has provided important clues to the relationship between shear in the radial electric field and the transition from the relatively poor L-mode confinement to the significantly improved H-mode confinement regime.

Future Prospects

There is a clear synergism between the need for improved diagnostic capabilities and advances in plasma science that either result from meeting that need or are the means to drive the improvement. Of the many areas that will continue to demand attention in the future, the most important may be the need to develop diagnostics for measuring properties of burning fusion plasmas. Most state-of-the-art diagnostic techniques have to be reexamined when the harsh neutron and radiation environment of a power-producing plasma is considered. With burning plasma, we are faced with the additional task of diagnosing the confinement and slowing down the alpha particles that must ultimately provide the power to sustain an ignited fusion plasma. Although much has already been accomplished in these areas, innovation in the next decade must proceed at a pace at least as rapid as that of the last 10 years. In addition, measurements of other fusion products, as well as stability and confinement of reacting plasmas in the presence of copious amounts of alpha particles, must proceed.

NON-TOKAMAK CONCEPTS

Introduction and Background

Fusion plasma physics has historically evolved through the exploration of a variety of magnetic configurations. The tokamak is the most highly developed concept, and most of the discussion presented above concentrated on this configuration. However, the need for innovative and diverse ideas is as vital as ever in view of the projected multidecade development that lies ahead for fusion.

The main non-tokamak concepts presently under investigation are the stellarator, the reversed-field pinch (RFP), and the compact torus. Each has potential advantages over the tokamak and is a unique source for new plasma physics information. Stellarators have the potential for a steady-state reactor without the need for inductively driven current. Reversed-field pinches, with a relatively weak magnetic field, offer the potential for a compact, high-power-density reactor with normal (non-superconducting) coils. Compact tori offer a reactor geometry in which the plasma torus does not link external conductors. The magnetic mirror approach to fusion is not discussed here since research on this concept ended in the mid 1980s.

The stellarator and reversed-field pinch share the same magnetic topology with the tokamak: toroidally nested, closed magnetic surfaces produced by helical magnetic fields. The relative strengths and origin of the toroidal and poloidal components of the magnetic field distinguish the different approaches. In the stellarator, the magnetic field is produced by current in external windings; in the tokamak, by plasma current for the poloidal field and an external current for the toroidal field; and in the RFP, predominantly by plasma current.

Recent Advances

Microwave and radio-frequency waves are now used to create "true" stellarator plasmas without a net internal current, thus avoiding the problem of "disruptions" that destroy plasma confinement. The absence of a net internal current has allowed detailed demonstrations of the nature of the pressure-driven "bootstrap" current (which must be maximized in tokamaks and minimized in stellarators for optimum performance) through its dependence on magnetic field curvature, as well as its control (e.g., reversing its direction). The plasmas are quiescent with no global instabilities. Operation in the "second stability" regime (in which plasma stability increases with increasing pressure) has been obtained and a connection made with energy confinement. External control has allowed a wide range of magnetic configurations. The similarities in plasma confinement between stellarators with very different magnetic configurations, and between stellarators and tokamaks, suggest that similar mechanisms may be responsible for global transport. The confinement scaling is similar in some tokamaks and stellarators, although stellarators exhibit a more favorable density dependence. The edge fluctuation levels, the corresponding particle transport, and the properties of the edge plasma are similar in both devices. Particle confinement is controlled by the edge properties. It has been increased by using electrically biased limiters and decreased by using magnetic error fields. The improved energy confinement behavior seen in tokamaks is now also seen in stellarators. Initial experiments with biased plates that intercept field lines exiting the plasma (divertors) are encouraging for eventual steady-state particle control in stellarators.

The RFP has evolved significantly during the past decade, both in the understanding of the physics and in the plasma parameters achieved. A fascinating property of the RFP is that it spontaneously generates a portion of its confining magnetic field. This constitutes a laboratory demonstration of the "dynamo" effect, analogous to the astrophysical dynamo responsible for magnetic field generation in stars. A thorough first-principles understanding of the dominant magnetic fluctuations in the RFP has been established through three-dimensional, nonlinear, magnetofluid computations. This theory agrees with experimental observation of fluctuation spectra and nonlinear three-wave coupling. It also offers a detailed explanation of the dynamo mechanism. The equilibrium magnetic field also is understood as a minimum energy state arising from plasma relaxation. These concepts carry over to tokamak phenomena, such as relaxation oscillations and disruptions. Recent attention has turned to the transport resulting from the fluctuations. Techniques have been developed to measure directly the energy and particle transport driven by magnetic fluctuations. As a result of the intrinsically low magnetic field in the RFP, all devices operate at high, reactor-level values of plasma beta. However, the quality of energy confinement, which is typically an order of magnitude worse than that in comparable-size tokamaks, requires improvement and is a topic of continuing research.

Another line of research is experiments with the so-called compact tori, with low aspect ratios. Stable equilibria have been produced in the laboratory, despite the absence of a strong toroidal magnetic field thought to be necessary for stability. This concept has the great advantage of a small unit size, which would significantly lower the cost of an eventual reactor based on this concept.

Future Prospects

Present experiments can develop much of the physics basis needed for improving stellarators, including tests of the contrasting optimization principles for the two main types of stellarators. The largest efforts involve complementary experiments in the United States, Japan, and Germany. Increased plasma heating power will permit fusion-reactor-relevant properties (higher pressure and improved confinement) in long-pulse (30-second) operation, optimization of the stellarator configuration and operational techniques for future large stellarators, and development of steady-state power and particle handling. Large superconducting-coil stellarators now under design and construction will allow true steady-state disruption-free plasmas without the need for externally driven currents or internally driven "bootstrap" currents. The higher heating powers available in these large stellarators will allow studies at higher plasma parameters (pressure, temperature, confinement time) needed for stellarator reactor development.

The evolution of the RFP as a fusion concept requires improvement in energy confinement. From the present experimental database, it is anticipated that transport will be reduced with plasma current. The highest RFP plasma currents operated to date are about 0.6 MA, for durations of less than 0.1 s. Currently, experiments are starting up in Italy that will produce 2-MA plasmas for 0.25 s. This will permit observation of the scaling of confinement on critical parameters, such as the Lundquist number (a measure of the electrical conductivity), which is particularly important for the MHD phenomena prevalent in the RFP. In addition, the evolving understanding of RFP fluctuations and transport is beginning to provide a scientific basis to develop methods to enhance energy confinement. Experiments are beginning in this area.

A major question in compact torus research is whether stable plasmas exist in which the ion radius of gyration about the magnetic field is small. Early experiments possibly were stabilized by the presence of ions with large gyroradii. Compact tori experiments with smaller gyroradii are just beginning.

CONCLUSIONS

The contribution of magnetic fusion research to the field of plasma science has been very significant. Besides being a driver for the development of modern plasma physics, fusion also has benefited greatly from the many advances in basic plasma science. Perhaps the most important area of future research is to learn how to "control" high-temperature plasma in modern confinement devices, which will require learning more about transport and devising effective means of controlling it. This also implies finding stable equilibria at the upper end of the high-beta regimes achieved to date and going beyond present beta limits, especially at high values of β_p (i.e., reduced plasma current in tokamaks). We must learn how to control radial plasma profiles, including those of temperature, density, and current density. At high currents, we must learn how to control disruptions, especially through current profile control. Control is clearly essential for achieving a more attractive fusion reactor based on the tokamak concept. In addition, pursuing other confinement concepts is important, particularly if attempts at control of the tokamak plasma fail or become too complex and expensive. It is also conceivable that a more effective confinement concept than the tokamak could emerge, especially if a steady-state reactor is desired because of technological considerations. However, in the past, funding limitations have often prevented a thorough development of alternate confinement concepts, with the possible exception of the stellerator.

In all confinement concepts, the issue of power and particle exhaust (helium removal) must find a solution in plasma science. This problem is just beginning to be addressed by the scientific community, and its solution will require a thorough theoretical analysis, often involving large codes, and experimental research in the area of "plasma edge" physics. To succeed, this study must include a combination of plasma science, atomic physics, and materials science. Finally, as the next generation of tokamaks enters the thermonuclear regime with burning D-T fuel, the generation of copious amounts of 3.5-MeV alpha particles will open the door to the study of alpha-particle-related plasma phenomena, including stability and transport. New diagnostics may have to be developed to study the interior of the burning plasma environment.

Unfortunately, in the past, many opportunities for fundamental scientific exploration were missed, in some instances because of funding constraints and in others because of changing priorities within the fusion program.

Perhaps the biggest problem in funding more scientific investigations in magnetic fusion is that the level of funding of this fusion program has decreased, in real dollars, during the past decade. Thus, painful choices have often had to be made between upgrading larger facilities to operate in high-performance regimes and increasing the scope of scientific investigations in intermediate-scale devices. Given the mission-oriented mandate of DOE's Office of Fusion Energy (OFE), further research and development will continue to shift toward issues related to burning plasmas, nuclear technology, reactor relevant materials, and so on. At the same time, other funding agencies, such as the National Science Foundation (NSF) and other offices in DOE, have not funded scientific investigations in high-temperature plasmas. If this trend continues, a serious void in the continued growth of high-temperature plasma science will result, despite its scientific merits.

The international fusion community is now engaged in the design of a major fusion facility, the International Thermonuclear Experimental Reactor. This facility will investigate the behavior of burning plasmas under conditions of intense self-heating by alpha particles, and it will demonstrate integration of the nuclear technologies required for a fusion reactor. In addition, the U.S. program has proposed construction of a national facility, the Tokamak Physics Experiment, to investigate modes of continuous operation under advanced performance conditions. The TPX, which is illustrated schematically in Figure 4.4, is planned to begin operation by the end of this decade, when it would become the "workhorse" for research in high-temperature plasma science in the United States.

RECOMMENDATIONS

For the continued development of plasma physics as a scientific discipline it is essential that there be a continued experimental capability to investigate high-temperature plasma phenomena. It is clear that a commitment to increased high-temperature plasma research and training of scientific manpower should be made now. With appropriate funding, the number of graduate students working toward a PhD in plasma-related fields is sufficient to meet such a commitment. The mainstay of this kind of research will remain the DOE Office of Fusion Energy (OFE). However, increased support for energy-relevant basic plasma science by the Office of Basic Energy Sciences (BES) at DOE, in cooperation with the OFE, which is recommended in Part I, would greatly benefit all energy-relevant plasma science and technology. This program could help fund specific experiments on large machines, as well as the operation of small and medium-sized experiments.

Funding at the level of several hundred thousand dollars per year per investigator would be of considerable value to university and industry efforts, even for participation in a large experiment. Initial investment in equipment at the level of a few hundred thousand dollars would also be of additional value. Diagnostic-type experiments could be carried out "piggyback" style at existing facilities.

Many plasma physics problems are best addressed in small- and medium-scale devices. Such devices can be used to test innovative confinement concepts, and the panel sees a need for two to three devices in the United States. In addition, somewhat larger-scale facilities would be desirable to continue basic research in high-temperature (a few keV) plasmas. Such devices might in some

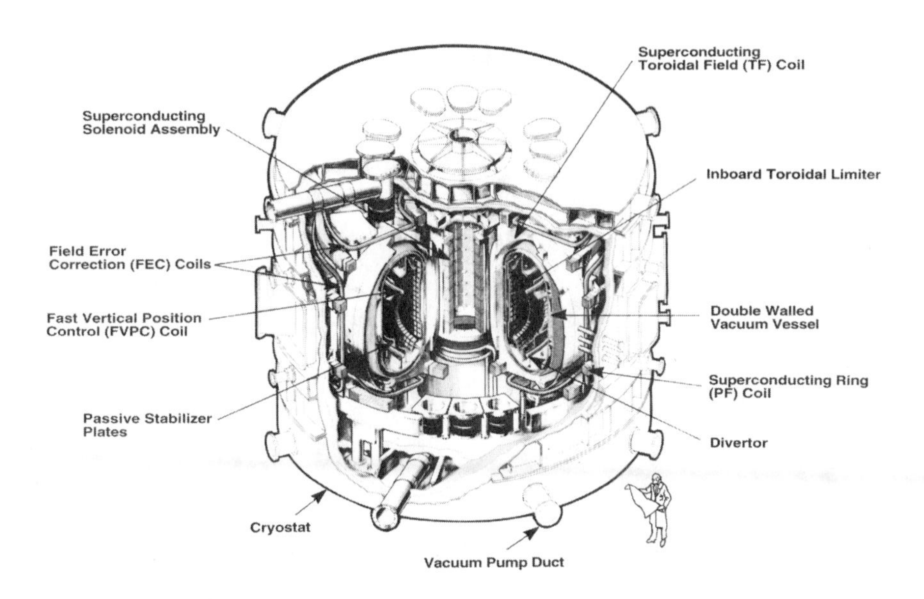

FIGURE 4.4 Schematic diagram of the advanced superconducting Tokamak Physics Experiment (TPX), proposed as a national facility to develop the scientific basis for a compact and continuously operating tokamak fusion reactor. Operating at a toroidal magnetic field of 40 kG and plasma currents of 2 MA, the TPX would investigate modes of enhanced plasma confinement and non-inductive current drive for plasmas lasting longer than 1000 s. (Courtesy of Princeton Plasma Physics Laboratory.)

cases function as user facilities, supported by a consortium of institutions and funding agencies. The panel envisions at least two such devices operating as user facilities. These facilities may be converted from currently operating and/or mothballed devices, most likely tokamaks. Appropriate nonohmic heating and current drive capability should be available on such a device.

Finally, it is also important to maintain a strong parallel program in theory and modeling, for it is the interaction between experiment and theory that facilitates the greatest progress in plasma science. In any case, it is clear that a commitment to continued support of research in high-temperature plasma science should be made now.

5

❖

Beams, Accelerators, and Coherent Radiation Sources

INTRODUCTION AND BACKGROUND

Consideration of the state and health of plasma science within the grouped disciplines of intense charged-particle beams, accelerators, and coherent radiation sources presents a picture perhaps representative of trends relevant to plasma science in general. Recent history suggests themes, several of which appear in common with other areas impacted directly or indirectly by plasma science. Basic and applied research have been supported indirectly within large Department of Defense (DOD) weapons-driven and DOE energy-driven application programs, such as the directed-energy weapons, nuclear weapon effects testing, and magnetic/inertial confinement fusion. There have been notable scientific and technical accomplishments in this area, along with visible examples of yet to be achieved or inflated expectations. Finally, and perhaps most important, there is concern about future funding availability and the organizational advocacy necessary to sustain, and advance, the underlying intellectual, facility and equipment infrastructure in light of evolving national defense, economic, and social priorities.

RECENT ADVANCES AND SCIENCE AND TECHNOLOGY OPPORTUNITIES

Intense Charged-Particle Beams

The mission for intense charged-particle beams has changed considerably over the last decade. Research and development sponsored by DOD, the Strategic Defense Initiative Organization (SDIO), and DOE resulted in facilities such as the Advanced Test Accelerator (ATA) at Lawrence Livermore National Laboratory for directed-energy weaponry, low-impedance multi-terawatt pulsed power machines for nuclear weapon effects simulation, and intense beams for fusion plasma heating. Kiloamp-MeV electron beams were developed that support high average power operation in excess of 100 kW with repetition rates approaching 1000 pulses per second (pps). Gyrotrons, devices that utilize a spinning electron beam in an axial magnetic field to produce millimeter waves for electron cyclotron resonance heating, successfully generated several hundred kilowatts in long pulses up to 3 s in duration at frequencies up to 140 GHz. Considering that 10 years ago, 100-ms outputs at 28 GHz and comparable power levels were representative, the technical community is justifiably proud of this technological accomplishment. Similarly, klystron technology has been advanced to higher frequencies (11.4 GHz) and powers (up to 100 MW), and the operation of a gyroklystron amplifier at the 20-MW power level at 11 GHz has been demonstrated.

Many of the military mission-oriented efforts have been canceled. However, industrial applications of high-energy electron beams, including bulk sterilization of medical products and food, toxic waste destruction via oxidation, and processing of advanced materials, are in the demonstration stage. Technology transfer from the laboratories to industry is being encouraged actively. Having invested several hundred million dollars over the past decade in developing intense charged-particle-beam systems for military use, the emphasis by federal agencies on technology transfer for industrial applications seems prudent. Charged-particle-beam parameters vary greatly, depending on the application. A NASA concept for beaming power to space requires basic plasma science research addressing such physics issues as low emittance growth ($< 20\pi$ mm-mr), beam breakup modes, and high current beam transport. Similarly, high-energy electron-beam systems proposed for toxic waste cleanup, enhanced welding, heat treatment, and material processing generally have less stringent requirements on voltage flatness and emittance, but require reliable generation and maintenance of very high average powers.

The interaction of intense charged-particle beams with plasmas, partially ionized gases, and matter offers rich scope for the study of strongly driven collective processes complementary to intense laser-plasma interactions. Electron and ion sources for intense beams have progressed from an empirical art to a developing science. Experiments, simulation, and analytical theory have contributed to this evolution, stimulated by the needs of inertial confinement fusion and other research programs.

Intense ion beams also make possible the creation of magnetic field-reversed ion rings in which the self-magnetic field of the circulating ion current exceeds the externally applied magnetic field. Such a ring would be a compact object of high energy density with unique theoretically predicted features, such as low-frequency stability. If these predictions are borne out experimentally, there may be many uses for such rings, including the possibility of ion acceleration to high energy for various applications. A state-of-the-art electron and ion accelerator is pictured in Figure 5.1.

Accelerators

Several new techniques have been demonstrated that can produce large-amplitude, coherent, high-phase-velocity electron plasma waves. These include the beat wave and wake field concepts. An electron beam accelerated by a beat wave accelerator is pictured in Figure 5.2. Such beat wave accelerators have achieved accelerated electron energies of 9 MeV within distances of 1.5 cm. Attaining 500-MeV energy gains at GeV-per-meter rates is considered a plausible goal within a five-year period in which progress, funding priority, and follow-on application potential may be assessed.

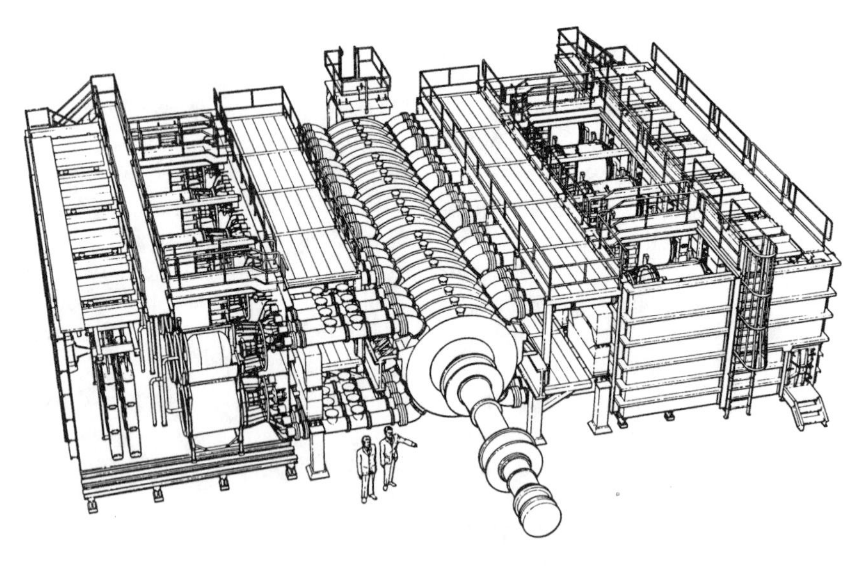

FIGURE 5.1 Schematic diagram of Hermes III, a 16-TW ion and electron accelerator that became operational in 1988. It represents a new class of accelerators that combine state-of-the-art pulsed power designs with high-power linear-induction accelerator cells and voltage addition along an extended magnetically self-insulated vacuum transmission line. This technology is particularly suited for applications requiring high output voltages (tens of megavolts), with megampere-level currents and short pulse widths (e.g., as small as tens of nanoseconds). Hermes III is used in its negative polarity configuration to generate an electron beam of ~20 MeV and 700 kA. It can also be configured in positive polarity to drive an ion beam diode. (Courtesy of J. Ramirez, Sandia National Laboratories.)

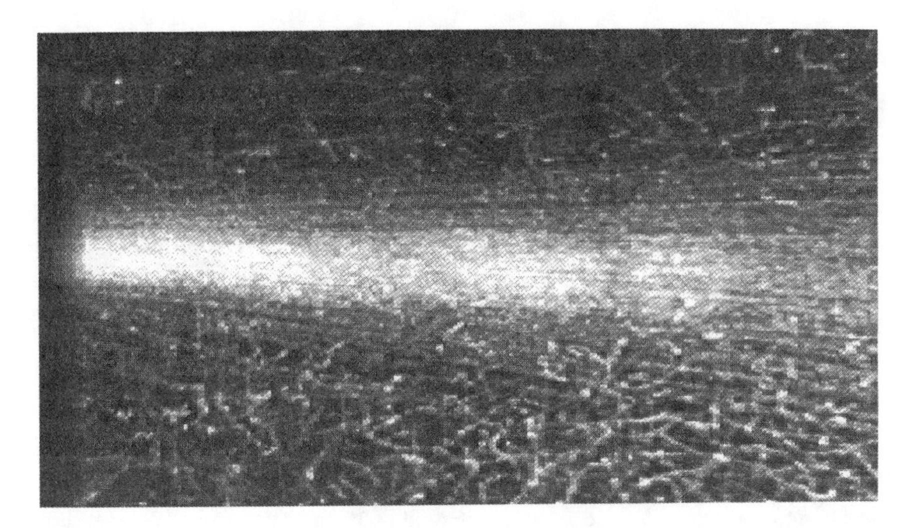

FIGURE 5.2 In a plasma beat-wave accelerator, a pair of laser beams fired into a dense plasma excite a high-phase-velocity plasma wave, and the electric field of this wave accelerates an externally injected electron beam. Shown is an electron beam that has been accelerated to more than 5 MeV in less than 1 cm in such a beat-wave accelerator. This technique holds promise for developing miniaturized particle accelerators for research, medicine, and industry. (Courtesy of C. Joshi, University of California, Los Angeles.)

These present and future very-high-energy, low-emittance, short-pulse electron beams should also further enable progress in other accelerator schemes, such as the plasma wake field accelerator. Relatedly, scaling principles for focusing electron and positron beams using thin plasma slabs as plasma lenses have recently been demonstrated, with 600-μm focal spot sizes achieved. In this case, basic plasma science is being exploited to develop an important "component" of an accelerator system rather than the entire system itself.

Relativistic 2 1/2-dimensional particle-in-cell codes, developed for inertial confinement fusion research, are now being employed to study the physics of short-pulse, ultrahigh-intensity laser-plasma interactions. Phenomena including severe hydrodynamic distortion by the intense light pressure, heating of electrons and ions to ultrahigh energies, relativistic penetration to supercritical densities, and relativistic self-focusing have been observed.

Other novel applications have also been developed, including frequency upshifting of electromagnetic radiation by reflection from ionization fronts and the generation of picosecond pulses of x-rays by irradiation of dense plasmas with ultrashort pulses. The wide-ranging progress may lead to compact particle accelerators, compact sources of tunable radiation, and new diagnostic tools for materials and biological applications.

DOE is supporting the development and operation of state-of-the-art "user test facilities" available to researchers in the field. Examples are the Accelerator Test Facility at Brookhaven National Laboratory and the wake field test facility at Argonne National Laboratory. On the other hand, opinions exist within the community that per-grant funding has not kept pace with what is now perceived to be required for conducting experimental plasma physics research, even with access to state-of-the-art facilities. This situation appears to be compounded further by the absence of any clearly identified "organizational champion" within NSF, DOE, or DOD.

Coherent Radiation Sources

As discussed above in Chapter 2, nonneutral plasmas, such as intense charged-particle beams, exhibit a wide range of collective phenomena. Some collective instabilities limit the performance of accelerators and storage rings and must be minimized; others can be used to convert beam kinetic energy into coherent radiation. New-generation coherent sources, which use electron beams and are based on beam instabilities, operate from the microwave range to the millimeter, infrared, visible and ultraviolet regimes, with previously unattainable intensities. The most prominent of these new systems is the free-electron laser (FEL). Other configurations include gyrotrons and cyclotron masers, and a variety of Cerenkov devices. The basic principle underlying these devices is electron bunching stimulated by an ambient, co-propagating electromagnetic wave. In a properly prepared system, the electrons of the beam, initially distributed at random, can be made to form clusters or bunches. If the bunch dimensions are comparable to or smaller than the wavelength of the desired radiation, coherent emission ensues. Thus, the principle of bunching is somewhat analogous to stimulated emission in conventional lasers.

Free-electron lasers have several important and distinctive features. The oscillation wavelength is not constrained to fixed transitions as in a conventional laser, thus allowing broad tunability. The pulse length is determined primarily by the electron beam, so that rf accelerators can be used without much difficulty to make picosecond pulses. Electron beams can transport high peak and high average power, making the FEL, with its reasonable conversion efficiency, a potentially attractive source of high-power radiation. Because there is no medium except the beam, problems associated with absorption and scattering can be avoided.

Over the last decade, pioneering studies have been carried out concerning the physics of the relevant nonlinear electron-wave interactions that govern the processes in these free-electron radiation generators. Concurrently, significant SDIO investment was made in free-electron laser R&D as a strategic missile defense system. Experimental, theoretical, and computational studies addressed relevant nonlinear interactions such as trapping and sidebands, mode-locking and phase stability, three-dimensional effects, time-dependent phenomena, and high-efficiency operation. High power (gigawatts) and high efficiency (40%) were demonstrated at the longer wavelengths. Systems have lased using storage rings, linear rf, induction and electrostatic accelerators, microtrons, and low-energy beams. However, the accomplishments within the strategic missile defense arena fell short of what was promised and expected, potentially leaving a blemish on an otherwise promising technology.

Most of the basic experiments referred to above were conducted on accelerators built for applications other than the free-electron laser. The SDIO sponsored several small-scale "user facilities." For the free-electron laser to find its appropriate place among coherent radiation sources, research is required to gain a detailed theoretical and experimental understanding of the temporal and spectral properties, to extend operation to shorter wavelengths in the VUV regime, and to increase the efficiency at the shorter wavelengths. A collaboration among academic institutions, national laboratories, and industrial organizations in the design and construction of a next generation of user facilities and the pursuit of the ensuing research would seem appropriate, given that the advocates and potential users of this coherent radiation source technology are successful in establishing its relative priority.

Coherent radiation source research in the x-ray portion of the electromagnetic spectrum includes synchrotrons/undulators, x-ray lasers, and harmonically converted short-pulse optical lasers. Of these three, major support has been given to the synchrotron/undulator effort (i.e., the Advanced Light Source at Lawrence Berkeley Laboratory and the Advanced Photon Source at Argonne National Laboratory). Brookhaven has an active users program dedicated to many areas of research, including biology, materials science, basic atomic physics and chemistry, and semiconductor physics. Much of the x-ray laser research work to date has been carried out with internal research and development funds at national laboratories. This review did not address the x-ray laser research conducted under the auspices of the Strategic Defense Initiative Office.

Concepts for soft x-ray lasers were developed successfully and demonstrated in the laboratory during the past decade. The generation of a dense, hot plasma by laser irradiation was a key feature of this success. This progress was the product of close collaboration among plasma theory, atomic physics, and laser science. Since the initial work, x-ray lasers at more than 50 different wavelengths have been demonstrated in about 10 laboratories worldwide. These x-ray lasers have been demonstrated with wavelengths from 400 to 35 Å, output powers to 100 MW, brightnesses eight orders of magnitude graeter than those of undulators, bandwidths of 5×10^{-5} at full power, and near-diffraction-limited and partially coherent output beam characteristics. Short-pulse harmonically produced x-rays are currently in the demonstration phase with wavelengths approaching 100 Å having been generated, albeit at low ($< 10^{-9}$) efficiency.

At present, synchrotrons have become undulators offering high coherent average power, but at great cost. The recent development of high-average-power glass lasers promises a high-average-power x-ray laser alternative to synchrotron sources at substantially reduced costs. Sources are envisioned with wavelengths between 100 and 150 Å and bandwidths of 5×10^{-5}, producing 10^{15} photons per pulse at 1 Hz, that will occupy physical footprints approximately 3 m by 6 m.

X-ray laser researchers aim at achieving shorter and shorter wavelengths and high coherent output energy. Exploiting ultrashort, subpicosecond laser pulses, photoionization pumping schemes render plausible lasing at or near 14 Å within 5 to 10 years. Harmonically generated x-rays from ultrashort-pulse laser-gas jet interactions are striving to achieve high efficiency at short wavelengths to go along with their inherent tunability and coherence. In turn, these sources can be envisioned to serve as drivers for further wavelength down-conversion providing improved radiation sources in the spectral regime of the order of tens of keV, which would significantly broaden potential medical and industrial applications.

The features of high power, narrow bandwidth, short pulse, and coherence make x-ray lasers attractive for future applications, such as biological microimaging, photoelectron spectroscopy, and probing of dense plasmas. (See Plate 4.) Given these scientific successes and potential applications and societal benefits, it is a serious deficiency that no federal agency has taken on the mandate to support x-ray laser research.

Advances in this area are hampered by the presence of large capital-intensive synchrotron and inertial confinement fusion facilities that historically have either siphoned off the majority of potential funding support or programmatically relegated the research to a piggyback status. The multidisciplinary nature of the research complicates the effort to accrue the critical mass of funding that would support a robust program.

CONCLUSIONS AND RECOMMENDATIONS

Research and development on particle beams, accelerators, and coherent radiation sources offers a wide range of opportunities for technological advances of importance for our society. Examples include food sterilization; waste treatment; welding and materials processing; advanced accelerator development; and the development of new, intense radiation sources for a wide range of applications. In the past, much of the basic science and development in this area was driven by military applications. However, given recent changes in emphasis on military needs, there is a danger that opportunities will be lost unless research and development continue to be pursued in areas in which there are significant technological opportunities. As discussed elsewhere in this report, there is a general need for support for small-scale basic research. This also is true in the areas of beams, accelerators, and radiation sources.

Given these considerations, the panel recommends the following:

1. Dual-use opportunities of defense-driven technologies should continue to be pursued.
2. Existing hardware and facilities no longer needed for ICF and SDIO applications should be made available for other scientific and technological applications.
3. Support should be given to small-scale basic research in these areas.
4. Particular attention should be given to the development of advanced concepts of particle accelerators and of new, intense radiation sources, such as x-ray lasers.
5. Where practical, the use of large facilities by outside users to pursue the scientific and technological goals described in this section should be encouraged.

6

❖

Space Plasmas

INTRODUCTION

Background

Space plasma physics is the study of natural plasmas in the solar system and associated technological applications. It is concerned with the ionospheric and magnetospheric plasmas of Earth and the other planets; the physics of the solar plasma internal to the Sun, the solar corona, and the solar wind; the interaction of the solar wind with planets, asteroids, dust, comets, and eventually the interstellar medium; and technology applications ranging from electric propulsion to space "weather" predictions. This is a vast, multiscale, physical domain wherein there are large variations in plasma sources, average thermal energy, flow velocities, magnetic field strength, and other physical parameters. The results are a rich collection of plasma physical behaviors that provide important intellectual challenges to fully understand the underlying processes. It is important to consider the wide variations in physical conditions in terms of spatial dimensions. The largest is that of magnetohydrodynamic flow phenomena. These include the massive outflow of solar plasma as the solar wind, which, entwined with the solar magnetic field, sweeps outward past the planets to its mixing with the interstellar medium. On a smaller scale, within planetary atmospheres, more complicated plasma flows are driven by electromagnetic fields associated with the interaction of the solar wind with the planetary body and atmospheric heating caused by solar radiation. On an even smaller scale, microscopic physical processes are occurring within plasmas in space. For example, the selective acceleration of electrons and ions in the high-latitude, high-altitude regions of Earth leads to the formation of highly structured streams of energetic charged particles that impact Earth's upper atmosphere, creating the aurora. Finally plasma technology devices such as electric propulsion and plasma contactors operate on a small size scale, establishing their own local boundary conditions and interacting with the nearby space plasma.

Knowledge of the physical processes operative in all of these examples is an important goal of space plasma physics for several reasons. First, it provides us with an understanding in quantitative terms of the variety of interrelated complex processes acting to shape and influence our terrestrial environment. Second, parts of the space plasma environment may be prototypical of the astrophysical environment. Third, space phenomena lead to fundamental scientific questions relating to the behavior of plasmas under conditions that can be very different from those created and studied in terrestrial laboratories. Finally, knowledge of the science underscores the development of technological applications operating in or based on the space plasma environment. As a consequence, investigations of natural space plasma processes extend the frontiers of human knowledge, enabling broader physical understanding of plasmas within the context of their general behavior.

Understanding Earth's plasma environment also has important practical consequences. Among these are an ability to model and predict ionospheric, magnetospheric, and interplanetary disturbances that could adversely affect ground-based communications, sensitive instrumentation in geosynchronous orbit, and the safety of astronauts participating in future interplanetary endeavors.

Status

The era of in situ exploration of space plasma physics began in 1946 with V-2 rocket "snapshots" of the terrestrial space environment and continues aggressively today. Measurement techniques include both direct sampling and space-based remote sensing. An excellent example of the latter is the global observation from space of aurora at UV and optical wavelengths, clearly delineating the dynamics of the auroral oval. The initial exploration of the terrestrial magnetosphere and ionosphere is now reasonably well complete, although there are still regions of the solar system that have not yet been explored at all (e.g., Pluto, the heliopause, the solar corona) and regions that have been seen only through brief flybys (e.g., Mercury, Uranus, Neptune). Emphasis now is shifting to the details of physical processes controlling these plasmas. The results of all modern theories and models have depended significantly on the progress of in situ observations.

Ground-based remote sensing studies of space plasma physics have played an important role by providing long-term, localized observations and understanding. Incoherent and coherent radar observations of natural ionospheric phenomena provide information with good temporal and altitude resolution from a few physical locations, thus complementing spacecraft measurements that give good global coverage. Magnetic and optical observatories help in elucidating global current systems and local energy deposition rates. Ground-based measurements that "modify" the natural plasma in the ionosphere have provided information on the physics of a variety of plasma instabilities and related phenomena.

Over the past two decades, our capability to numerically investigate the behavior of space plasmas has steadily improved. Models and simulations of the 1960s and 1970s evolved as a consequence of attempts to understand particular features of the solar-terrestrial environment (e.g., the composition and thermal structure of the atmosphere and ionosphere, the dynamics of interhemispheric plasma interchange, the coupled dynamics of energetic plasma in the magnetosphere and electric fields and currents in the magnetosphere and ionosphere, the interaction of the solar wind with the geomagnetic field, the formation of shocks in the solar wind, the propagation of solar and galactic cosmic rays in the solar wind, and the dynamics of magnetic reconnection). More recently, the ability to study plasmas on a microscopic scale has evolved through the use of various simulation techniques with supercomputers. These codes permit the investigation of various modes of plasma dynamics associated with internal energy and momentum transfer between the plasma constituents and plasma waves. Unfortunately, owing to limitations of computer resources, these studies are often limited in terms of their spatial and temporal resolutions.

The last 35 years of satellite exploration and ground-based experiments, going hand in hand with theoretical modeling and simulation, have put us at a stage where the gross plasma morphology of the solar system is defined in an average sense. This large-scale picture is a synthesis of a relatively few observations that are localized and scattered in both space and time. The major task ahead in our studies of space plasma physics is to obtain the necessary information to be able to understand and elucidate the processes that control the behavior of these plasmas. This will require the use of sophisticated, multispacecraft missions, accomplishing direct and remote sensing observations as well as active perturbation experiments. Ground observations and experimentation will continue to provide important long term measurements. Both space-based and ground experimentation will have to be coordinated effectively. Advanced computational techniques will dramatically strengthen theoretical modeling and simulation.

TOOLS FOR SPACE PLASMA PHYSICS

Space-Based Techniques

The mainstay of progress in space plasma physics has been in situ and remote sensing experiments from space. They provide the means for systematically monitoring large regions of space plasma. The inability to distinguish space from time changes in measurements from a single spacecraft underlies the future thrust toward flying constellations of identically instrumented and electronically coordinated satellites to study a given phenomenon or a limited region of space. The success of such efforts will depend strongly on the implementation of self-contained "smart" electronics to facilitate real-time complementarity and software techniques for onboard selection, digestion, and compaction of the plethora of data from multiple sources.

Active experiment techniques are used to create a controlled disturbance and study its effect on the environment. (See Figure 6.1.) Active experiments have a broad range of objectives. These include (1) simulation of natural processes occurring in space plasmas, (2) measurement of physical properties such as reaction rates of atmospheric constituents and collisional cross sections, (3) use of space as a laboratory without walls to study fundamental plasma physics, (4) probing the natural environment as is done in experiments tracing magnetic field lines by electron beams, and (5) improving communication systems by studying the propagation of electromagnetic waves. For the study of space plasma processes the attraction of active techniques is twofold: there is no need to wait for a phenomenon to occur naturally, and the source characteristics are known and can be controlled. In this way, active experiments are similar to laboratory experiments except that the former have the advantage that the space plasma for most purposes can be considered boundless. The disadvantage is that it is difficult to obtain measurements with high spatial resolution. To remedy this problem, multiplatform experiments have become more common in recent years.

Ground-Based Techniques

Although the magnetosphere is an enormous region in space, we benefit largely from the dipolar origin of the field, which causes all the geomagnetic field lines of the magnetosphere to intersect the Earth, and most of them in the polar regions. This focusing of field lines provides a tremendous benefit observationally because arrays of ground-based instrumentation are relatively inexpensive to deploy and operate and they can provide important correlative data as well as a global context within which to interpret satellite data.

Ground-based data also provide a long-term database that permits understanding of the secular variations and changes in the solar-terrestrial "climate."

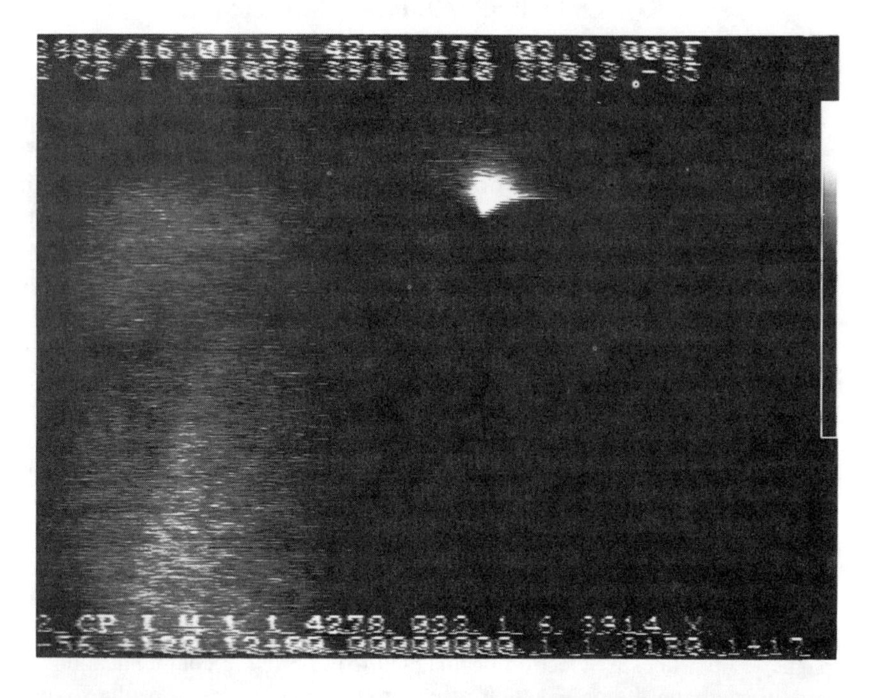

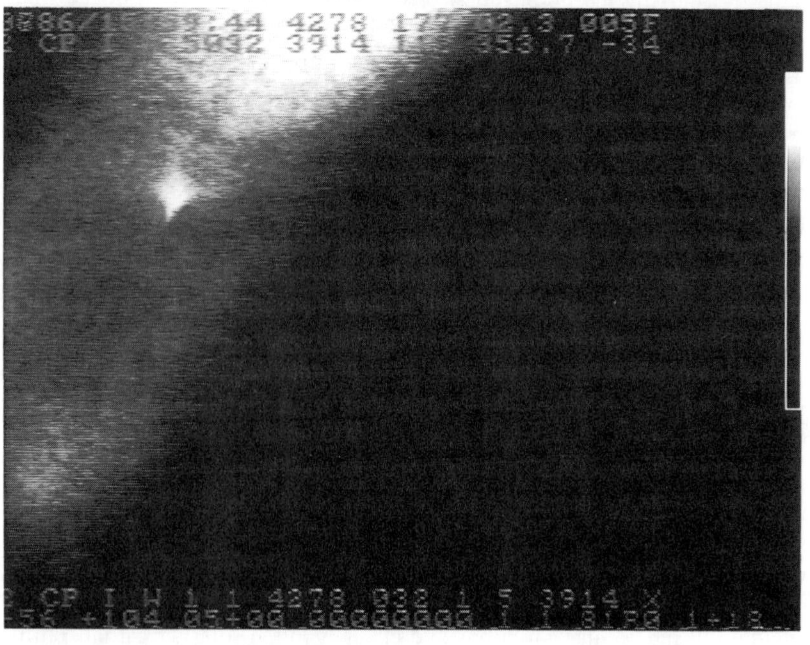

A variety of standard observatories provide continuous measurements of magnetic fields, ionospheric conditions, and solar activity. Also, a number of facilities operate only occasionally, such as the large incoherent-scatter radar facilities.

Plasma Theory and Simulations

Theory must develop a framework for interpreting observations of various physical systems. With this framework as a basis, quantitative analytic and numerical descriptions of model physical systems are constructed and refined through comparison with observations. Ultimately, the model physical systems must be sufficiently similar to the natural physical systems, and the numerical and/or analytic descriptions of these model systems must be sufficiently refined to provide a high level of predictability of the observed behavior of space plasmas. (See Plate 5.)

An outstanding issue involving modeling and simulations is how to properly represent multiscale phenomena numerically. For example, how can one best incorporate anomalous transport processes occurring in boundary layers (such as shocks), which occur on both temporal and spatial scales that are microscopic, into meso- or macroscale numerical codes?

The ultimate test of accomplishment in theoretical and simulation research must be the degree to which closure is achieved with ground and space experiments. Theoretical investigations may be conducted with the goal of explaining physical phenomena that have been observed and measured, and theoretical studies may be designed to lead to predictions testable in space plasma physics missions.

FIGURE 6.1 Example of an active space-plasma experiment designed to study the deposition of energetic electron fluxes in the atmosphere. This controlled experiment, which was flown on the Space Shuttle ATLAS-1 mission in 1992, used an electron beam and an optical imager to study the deposition of high-energy electrons into the polar atmosphere. Shown in the top panel is the artificial aurora generated by the beam (upper right in the figure) and a quiet auroral arc (left). The electron beam pulse was 1 s in duration. The camera viewed downward along the magnetic field direction, and the direction of motion of the Shuttle Orbiter was to the left. The width of the image at a height of 110 km is about 80 km. The bar is linear in optical intensity. The image in the bottom panel is the same as that in the top panel but taken at a later time. It shows the artificial aurora superimposed on a large auroral arc. (Courtesy of S. Mende, Lockheed Research.)

Laboratory Techniques

Historically, laboratory experiments related to space phenomena never have been supported significantly by funding agencies, the exception being specific technology efforts such as electric propulsion. One reason is that it is costly and difficult to build experiments wherein the magnitudes of critical parameters scale with their space counterparts, a necessary condition for the laboratory work to be relevant. However, as a result of advances in general technology (i.e., developments in computers, digitizers, and other hardware), as well as large strides in the design and improvement of plasma devices, it is now possible to perform experiments that were not possible 15 years ago. This increased sophistication allows the possibility of meaningful laboratory simulations. Laboratory experiments can probe a process with unprecedented detail and uncover effects that may not be easily detectable in space. (See Plate 6.)

Laboratory experiments can address both local and global physics issues, the latter often determined by boundaries. In some cases, one can comprehensively analyze physical phenomena simultaneously from both global and local points of view. Furthermore, experimental devices may be rapidly configured to perform new experiments as ideas are developed. This can happen on the time scale of days or weeks. The hardware is reusable and flexible. Many different experiments can be performed on the same machine. In addition to processes directly related to space plasmas, laboratory experiments have provided important technological advances within the areas of spacecraft propulsion and spacecraft potential control. Propulsion devices such as ion thrusters and arcjets, and plasma bridges, such as the plasma contactor, are being tested and studied in space with diagnostics developed largely for ambient plasma observations.

FUNDAMENTAL PROCESSES IN SPACE PLASMAS

Summarized below are a few basic phenomena of wide significance.

Wave-Particle Interactions

The important role of plasma waves in the macroscopic transfer of energy and momentum in space plasmas has become clear as a result of complementary and strongly coupled experimental and theoretical investigations. Over the last few decades, spacecraft observations have provided valuable information regarding the plasma environments around the Earth, planets, Sun, and comets. For example, from a limited set of planetary wave observations, it has become clear that similar waves exist around all the magnetized planets. This would seem to set limits on the significance of anthropogenic effects in triggering natural waves in Earth's magnetosphere. By far the largest body of information on the role of waves has been accumulated in Earth's plasma environment. Spacecraft observations have characterized Earth's magnetosphere as a collection of discrete regions with distinctive physical properties separated by well-defined boundaries. Strong plasma waves have been observed within the boundary regions, particularly the magnetopause and plasma sheet boundary layers. For the magnetopause boundary layer, this has important consequences for the entry of solar wind plasma into Earth's magnetosphere. Spacecraft have mapped out the locale and statistics of occurrence of most important classes of plasma waves in Earth's magnetosphere. Besides being important for boundary region dynamics, plasma waves provide the dominant loss mechanism for energetic electrons in the inner magnetosphere via pitch-angle diffusion and are drivers for the precipitation of ions and electrons into the lower atmospheric regions. Plasma waves are thought to play a principal role in heating ionospheric ions in the topside ionosphere. Upwelling of these heated ions along magnetic field lines and their subsequent trapping in the equatorial plane due to interactions with regions of plasma turbulence provide an important source of magnetospheric plasma. A wide variety of plasma waves participates in the complex energy transfer processes on auroral field lines. Plasma waves have been used as a diagnostic tool to obtain properties of the plasma from both ground-based and space-based systems. Plasma irregularities in the ionosphere occur with scale-size distributions covering tens of kilometers to fractions of a meter. These irregularities, which result from poorly understood instability mechanisms, are a major source for the disruption of high-frequency (HF) and extremely-high-frequency (EHF) communication systems.

Charged-Particle and Plasma Energization

Charged particles in plasma can be accelerated to high energies through a variety of mechanisms, some of which occur in nature or can be induced in suitably arranged space experiments. These include particle acceleration through resonance with quasi-monochromatic waves; stochastic acceleration resulting from resonance overlap due to large wave amplitudes or the presence of a finite spectrum of waves; acceleration by parametric processes, such as beat waves, Brillouin, and Raman scattering; and acceleration by electric fields that result from changes in macroscopic plasma morphology as encountered, for example, on auroral field lines. These phenomena are fundamentally nonlinear and extremely complicated, from both theoretical and observational points of view. Although measurements of electric and magnetic fields can be made with very high time (spectral) resolution, particle measurements are comparatively crude. For many purposes, the particle distribution function must be known to an accuracy that cannot be obtained with present-day technology. As an example, only two of the three velocity components of a distribution are generally known (perpendicular and parallel to Earth's magnetic field). Yet in many resonance interactions with waves, it is the unknown component (the phase coherence) that is the key to the interaction.

Dust-Plasma Interactions

Dusty plasmas are the most common type of plasmas in space. It is now believed that the long-term evolution of the dust and plasma environments is strongly coupled. The dust grains collect electrostatic charges from the plasma, and the evolution of their spatial distribution, size distribution, and lifetime can be determined by electrostatic forces and plasma drag. On the other hand, the dust can alter the plasma composition, density, momentum, and energy distribution, as well as the dispersion relations of the waves propagating in a dusty plasma medium.

In the past decade, a growing effort (laboratory experiments and theory) has focused on problems related to dusty plasmas. We now understand the important processes that determine the charge of the dust grains and have learned the transport processes that shape the fine dust components in planetary rings embedded in magnetospheric plasmas. Magnetospheric perturbations were clearly shown to be responsible for the observed spatial distribution of small dust grains in the Jovian and Saturnian rings. Collective dusty plasma effects were suggested to explain the spokes (transient radial dust features on Saturn's main ring system) observed on Voyager images. The large scattering cross section of charged ice grains in noctilucent clouds is thought to be responsible for the observed anomalous radar echoes. The differential settling of bigger and smaller grains toward the midplane in the early solar system was suggested to cause spatial charge separation that might have resulted in large-scale electrostatic discharges. These lightning bolts could explain the existence of chondrules (small molten beads of rocks found in meteorites).

The Critical Ionization Velocity Effect

Investigations of the critical ionization velocity effect are an important part of space plasma science. The phenomenon involves the nonclassical ionization of energetic neutral atoms and molecules as they move through a background magnetized plasma. From laboratory studies and some space measurements, it is thought that when the center of mass energy of the neutrals rises above their ionization threshold, there is rapid ionization of the neutrals. This process apparently involves energization of the ambient electron gas by plasma waves associated initially with the transformation of a few energetic neutrals to ions. The newly born ions have considerable kinetic energy and heat the electrons through collective plasma processes. When sufficient neutrals are converted to ions, as might happen through charge exchange, for example, the energy density of the electron gas rises to the point where additional ionization of the neutrals ensues, and a flash ionization of most neutrals occurs.

This phenomenon has great significance for models of young solar systems. To perform comprehensive measurements of the processes involved, it will be necessary to achieve the correct physical scale: the electron gas must be heated over a sufficiently large distance that its temperature can rise to the point where impact ionization of the neutrals becomes important to the overall system of interacting gases. Such experiments lie in the future and will require much more extensive supporting resources than have been possible with small free-flying satellites or rockets.

Radiation Processes

The topic of radiation processes is relatively new and involves detailed study of the production, transport, and absorption of microwave, infrared, and shorter-wavelength radiation in dense plasmas. However, its implication to the study of astrophysical systems is profound. The interaction of such radiation with matter involves individual molecules, atoms/ions, or electrons—not collective plasma processes. Clearly, radiation processes are of fundamental importance in transporting energy through portions of the Sun and of the Earth's atmosphere. In addition, radiation propagating freely from its source and from optically thick regions is the primary means by which remote sensing is accomplished. The opportunity to study fully coupled electromagnetic radiation with plasma dynamics in the space environment supplements the extensive work done in laboratory plasmas on similar problems.

ACTIVE EXPERIMENTS

Active experiments have a broad range of objectives. The techniques used in active experiments include four main categories: (1) injection of plasma and neutral vapor; (2) injection of energetic beams of neutral particles, ions, or electrons; (3) wave injection from ground based systems of acoustic waves and electromagnetic waves in the very-low-frequency (VLF) and HF bands, or injection from space vehicles of VLF, HF, and microwave radiation; and (4) use of the spacecraft as a disturbance to study spacecraft wake, vehicle charging, ram glow, or the electromagnetic effects of tethered systems.

Plasma and Neutral Mass Injections

The natural space environment can be modified by the introduction of foreign gases and plasmas to induce or enhance local processes. These include changes of the local ion composition, reduction of the local electron density, changes in the charge state of ions, changes in the average energy of the local plasma, and the plasma's chemical nature. This permits study of various processes occurring within ionospheric plasmas, including production, loss, and transport. It is also a way of creating unstable environments that evolve in interesting and new ways not normally found in the natural environment. The possibility of creating a large-scale ionic plasma (positive and negative ions dominating the overall composition) is both interesting and important in that it allows new processes to become dominant in plasma behavior. Such experiments are difficult, if not impossible, to perform in terrestrial laboratories.

Pulsed plasma beam or contactor experiments, where a dense plasma cloud is released into the ambient medium, can give information on plasma transport. The plasma cloud expands and distorts in response to its internal diamagnetic structure as well as the external flow field. This expansion sheds light on the fundamental plasma physics of high-beta plasma clouds, such as occur in the magnetotail, as well as the nature of the transport process when the cloud is diluted. Such experiments are conceptually similar to those already under way in ground-based laboratories, with an important exception. By using pulses of sufficient density and duration, it is possible to create steady-state diamagnetic plasma regions near the source. Information about the various processes acting in such an unusual plasma configuration is an important step toward understanding a new regime of plasma physics.

Particle Beam Experiments

Particle-beam experiments have been conducted from many sounding rockets, satellites, and space shuttle missions. Objectives in these experiments have been (1) to map the geomagnetic field structure and parallel electric fields by observing echoes of beam electrons from the magnetic conjugate mirror point or from electrostatic structures along auroral field lines; (2) to study auroral processes, such as optical emissions and wave turbulence in auroral particle beams; (3) to stimulate electromagnetic and electrostatic wave excitation; (4) to create suprathermal electron tails; (5) to observe the interaction of the particle beam with the neutral gas in the vicinity of the source payload; and (6) to study spacecraft charging and neutralization.

Wave Injection Experiments

Space-based wave injection experiments make use of a number of different techniques to launch waves into the plasma. Topside sounders rely on the excitation of plasma resonances. Transmitters have been used on a number of sounding rockets. Finally, modulated electron beams have been used as "virtual antennas" on STS-1 and the Spacelab-2 shuttle missions and on numerous sounding rocket experiments. Waves have been detected to a distance of a few kilometers. In these experiments, receivers have been located on the transmitter platform, on a subsatellite, or on the ground. Objectives include the study of antenna properties, near-zone electromagnetic field studies, wave propagation, and wave-particle interactions.

Ground-based HF wave (megahertz-range) injections are launched from powerful radars into the ionosphere under quiet conditions, during magnetic storms, or in conjunction with chemical releases from active experiments. A large number of plasma processes can be studied in this way: focusing or defocusing of the beam, interference with communications, heating of the plasma, generation of suprathermal electron fluxes and airglow, instabilities including self-focusing, parametric interactions and strong Langmuir turbulence, generation of plasma irregularities, focusing by chemical releases, and effects of a hierarchy of heater thresholds. In addition, pulsed HF heating is used to modulate the auroral electrojet current for extremely low frequency (ELF) or VLF wave generation. VLF wave experiments study wave-induced particle precipitation, earth-ionosphere wave guide modification, stimulation of VLF emissions, direct D-region heating, and ELF-modulated VLF to produce a polar electrojet antenna. Strong acoustic waves from explosions generate gravity waves and acoustic shock waves that couple into the plasma through collisional interaction.

Vehicle-Environment Interactions

A space vehicle perturbs the environment in a number of ways. Out-gas clouds, fluid dumps, and thruster firings interact with the ambient plasma much as the neutral gas injections described above. In addition, the structure itself creates a wake in the plasma, in particular for orbiting platforms, for which the spacecraft velocity generally is larger than the ion thermal velocity. Surface glow induced by neutral atmosphere interactions with the spacecraft surface has been studied in the case of the space shuttle. High-voltage power systems and their interaction with the ionosphere were studied in sounding rocket experiments. Objectives were to study the plasma sheath, the charging levels, and the steady-state currents in the ambient plasma.

Processes associated with the physical contact between plasmas and exposed surfaces in space are an important practical aspect of many advanced scientific and technological space systems. For example, the ability to draw electron current from magnetized space plasmas is an essential feature of plans for power-producing electrodynamic tether systems. Charging of dielectrics in the vicinity of high-current beam experiments is similarly an important concern. It is striking to realize that while basic issues of plasma sheaths and current extraction have been known for more than 50 years, we still lack fundamental knowledge of the processes involved, especially at high voltages and currents. Relatively simple experiments, such as measuring the voltage-current collection curves for magnetized plasma, have yet to be done for ranges of parameters in which large-amplitude plasma waves play an important role.

Much of our present knowledge of plasma sheaths comes from laboratory measurements. In the case of electron current collection, this has imposed severe limitation on the scale of phenomena that can be studied. Because the total number of electrons in a given plasma chamber is limited, laboratory measurements of electron current are limited in time and current density to very small values. Measurements in space offer a far better situation since in space one can place the collecting anode in an essentially unbounded medium.

FUTURE PLANS AND OPPORTUNITIES

In Situ Observations

The major task ahead in our studies of naturally occurring space plasmas is to obtain the information necessary to understand and elucidate processes that control their physical behavior. This will require the use of sophisticated, multi-spacecraft missions containing the latest technology in direct and remote sensing instrumentation. The technology to carry out these studies exists today. We now know how to construct rugged, reliable, and lightweight instruments capable of making three-dimensional, high spatial and temporal measurements of particle fluxes. We have demonstrated that we can make detailed measurements of electric and magnetic wave phenomena and have had great success in making remote optical measurements from UV to microwave frequencies.

There are plans to make some of these important measurements in the decade ahead, using relatively small, as well as larger, spacecraft programs. Some focused studies can be carried out with a single low-cost spacecraft. An example of such a mission is the Fast Auroral Snapshot (FAST) Explorer. It is a relatively low-cost mission planned to be launched in 1995. Its aim is to study the plasma microphysics of the terrestrial auroral zone, and it will make very-high-resolution measurements triggered by certain preprogrammed signatures.

The possibility of even cheaper, but still highly sophisticated missions, using leading-edge civilian and military dual-use technology, is currently under study. The microelectronics revolution has enabled the design of small, fast, smart, less expensive instruments, compared with the standards of a decade ago. Microprocessors, use of higher-order software languages such as C++, and specialized semiconductor chips, which enable digital signal processing, analog-to-digital signal conversion, and other operations, can be built into instruments, providing wide flexibility and speeding up changes in operating modes and other functions. Specialized analog systems under digital control permit rapid and accurate changes in voltages, currents, and other important aspects of instrument operation. As a consequence, aperture sizes have shrunk toward theoretical limits, detector systems have become miniaturized, detector efficiencies have become high, power consumption has become very low, and data rates are fast enough to challenge every satellite or suborbital rocket system designer.

Furthermore, the development of microprocessor systems capable of controlling all aspects of remote experiments has opened the way to new concepts of experiments, enabled by high data bandwidths and precise timing of the process or events under consideration. The combination of rapid switching of instrument modes, linked to high bandwidth data acquisition, and the ability to analyze data onboard the instrument platform and alter the course of the data taking, is a feature that is not fully in place but that opens the way to much more precise observations of plasma phenomena in space.

However, many, if not most, of the major outstanding questions in space plasma physics require sophisticated and coordinated multiple satellite missions to provide the much needed ability to distinguish between spatial and temporal changes. This is not a new concept. The Global Geospace Science (GGS) Program is a part of the International Solar-Terrestrial Physics Program (ISTP) and consists of three satellites—Geotail, launched in 1992, Wind, launched in 1994, and Polar, to be launched in 1995. The planned separation distance of these satellites is very large; thus, their mission is to study long-range correlations. Cluster, a European Space Agency (ESA) program, planned to be launched in 1995, is the first constellation mission. Four essentially identically instrumented satellites are planned to fly in a tetrahedral formation, with variable separation, which at times will be as small as a few ion Larmor radii. Another mission in the study phase, the Grand Tour Cluster (GTC), is aimed at studying low-latitude magnetospheric structures even smaller than an ion Larmor radius.

All recent large-scale programs, such as GGS and Cluster, have involved international cooperation. Such joint endeavors are both necessary and desirable. Involvement of international partners not only decreases the cost involved for the participating nations, but also ensures that the best available technology is used and the best scientists are participating, thus enhancing the scientific return. Careful coordination of ground-based observations with satellite-based measurements also will lead to significant increases in the scientific return from such programs.

Solar physics in general, and solar plasma physics in particular, are different in the sense that our knowledge to the present has been obtained largely from Earth or Earth orbit. The perspective—but not the proximity—changed in the fall of 1994 with the passage of the Ulysses spacecraft high above the Sun's south pole at a radial distance of about 2 au (where 1 au is the Sun-Earth distance). Despite the resultant limitations of poor spatial resolution, we infer strongly that a rich variety of plasma phenomena are occurring and that plasma physics is complementary to nuclear physics in determining solar structure and behavior. The Sun is a source of magnetic field, which probably is generated by the interaction between its differential rotation and MHD convection in its interior. The body of the Sun supports a plethora of waves, which manifest themselves through surface oscillations, whose study has given rise to the field of helioseismology. The magnetic field reaching the surface, rather than being uniformly distributed, is concentrated in small flux tubes, many having loop topology, so that the solar corona, as viewed in x-rays (see Plate 7), is a veritable archive of plasma structures: sunspots, fibrils, prominences, spicules, holes, bright spots, and so on. The corona is a dynamic region, the source of both a continuous solar wind and, from time to time, blobs of localized, energetic plasma flow that drive shock waves ahead of them as they propagate outward toward the planets. Such coronal mass ejections (CMEs) usually cause ground-based electromagnetic disturbances when they hit the terrestrial magnetosphere. Solar flares are associated with many CMEs and are one of nature's most observable examples of particle acceleration. Radiation spanning the electromagnetic spectrum is generated either directly by wave-particle processes or secondarily by energetic particles interacting with chromospheric material. In addition, flare-related relativistic electron beams propagating into the solar wind produce there characteristic (Type III) radio waves by processes that are generally thought to be highly nonlinear in nature.

Understanding of plasma activity on the Sun would undoubtedly prosper from a high-resolution, FAST-type mission. That is impossible because of the distant and more hostile environment. The closest approximation is the Solar Probe spacecraft, currently under study, which would make a one-time pass within three to four radii of the nominal surface. Besides providing scientific insight through in situ observations, Solar Probe presents obvious technical challenges in the area of thermal engineering.

In the meantime, the ESA Solar Optical and Heliospheric Observatory (SOHO) is being prepared for a 1995 launch, with U.S. participation in the instrument complement. The scientific objectives of SOHO are to study localized plasma structures—loops, prominences, holes, flares, mass ejections, and so on—in the solar chromosphere, transition region, and corona by spectroscopy and imagery of their electromagnetic emissions at UV and visible wavelengths and, at the same time, to monitor derivative solar wind effects via onboard particle measurements. Additionally, data from instruments that measure fluctuations in solar brightness and coherent, long-wavelength oscillations of the solar disk, so called helioseismology, may shed light on processes occurring in the solar interior.

Cassini, another mission in an advanced state of development, is currently being developed as a mission to Saturn scheduled for launch in 1997. Using a new technique, one instrument will be able to form a two-dimensional image, providing the direction of arrival of energetic neutral atoms formed via charge exchange with energetic ions. The results will enable the measurement of the spatial extent and energy composition of the large plasma zones surrounding Saturn and its satellites.

Much future work in planetary science will focus on waves and instabilities in naturally occurring dusty plasmas. The Ulysses, Galileo, and Cassini missions will fuel more interest in this field. Data from dust detectors, imaging, plasma and plasma wave experiments, magnetic field measurements, and so on will be used to understand dusty plasmas in planetary magnetospheres and in the interplanetary medium.

An important new mission to study the ionized and neutral upper atmosphere of Earth is slated for a new start within the next year. The Thermosphere-Ionosphere-Mesosphere Dynamics (TIMED) mission will carry a variety of instruments designed to probe complex interactions affecting the behavior of Earth's atmospheric regions lying above the stratosphere. The upper atmosphere has a factor of 10 greater response to global warming than the lower atmosphere, and can serve as an indicator of subtle changes that may be either anthropogenic or externally driven. TIMED will study this altitude regime, which experiences coupling between neutral and plasma constituents, and where competition between solar irradiance variation and plasma processes such as joule heating is important. Plans for deployment of an Earth Observing System as part of NASA's Mission to Planet Earth should incorporate as a component of the program the study of plasma coupling to the neutral atmosphere.

A major difficulty with space satellite constellation experiments is that differences in satellite altitudes lead to different orbital periods. Coordinated local observations thus become a matter of occasional opportunity, and a concentration of observations from different satellites at one time will rapidly decay to widely dispersed observations over times of a few minutes. NASA is developing a new way to obtain coordinated measurements over distances up to several hundred kilometers. This involves the use of long tethers connecting individual satellite platforms together. In a static configuration, the instrument string (looking much like a deep-sea string of acoustical sensors) is deployed along a vertical direction. The entire system moves with a constant, common angular velocity with respect to the Earth. This results in the possibility of obtaining simultaneous plasma and atmospheric data over a wide range of altitudes. Such a system could be used to observe the high-altitude acceleration zone for auroral electrons, the possible presence of horizontal plasma shear in large-scale plasma convection in the polar caps, or the behavior of aurora plasma in the regions of atmospheric excitation.

A substantial number of other missions exploring the behavior of space plasmas are now being planned by the U.S. and international scientific communities, with the Solar-Terrestrial Energy Program (STEP) providing coordination of both ground- and space-based systems. As in the past, there is a strong sense of cooperation among the international participants. An extensive and specific evaluation of future space missions, to be entitled *A Science Strategy for Space Physics*, is currently in progress under the auspices of the NRC's Committee on Solar and Space Physics and Committee on Solar-Terrestrial Research. When completed, this study will be used by NASA in its planning of new missions over the coming decade.

In Situ Experiments

The previous section concentrated on passive observations of natural processes acting in space. It is also possible to conduct in situ, or active, experiments whereby artificial injections of charged particles, neutral gases, or electromagnetic waves are used to alter natural processes or to stimulate new processes in the ambient plasma medium. It is possible to contemplate a rich selection of potential in situ experiments capable of exploring new areas of plasmas in space. NASA is the principal sponsor of such work, but for the past four years, as a matter of policy, NASA has restricted its funding to those projects that explore natural processes, rather than artificially induced behaviors. However, since many of the results of the latter types of experiments have important implications for plasmas in different space environments, it is hoped that this policy will be reviewed. Here we give brief outlines of some possible in situ experiments that have special merit.

Space vehicles offer the promise of performing three-dimensional experiments in unbounded plasmas with varying mixtures of neutral gas. These can be done on a scale size that should make the instrumentation easy to build. In addition, the relevant time scales are microseconds or longer, which are easily measured and recorded. In spite of these advantages, plasma experiments in space have not been easy to perform. The principal reasons are that diagnostic instruments are difficult to place accurately and the space platforms that carry them may be big enough to interfere with the experiment.

By using space platforms with suitable resources, it should be possible to investigate steady-state diamagnetic cavities in space plasmas. In this situation, the plasma effusion speed from its source can be made larger than the diffusion speed of the magnetic field. A complex region of low magnetic field is maintained by plasma pressure against the flowing ambient plasma and ambient magnetic field. This is an unstable situation, which opens the way to investigation of various types of instabilities. It is likely that these will reveal the presence of many new high-beta plasma-magnetic field interactions that depend on various plasma and magnetic field parameters. Magnetic field interactions, analogous to the solar wind-geomagnetic field coupling, can also be anticipated as the capability to construct and operate large magnets in space evolves. These experiments, involving a variety of plasmas and magnetic field configurations, will have relevance to a wide range of astrophysical situations.

Terrestrial Observation Networks

Support for the existing standard observatories, which provide the long-term monitoring of fundamental parameters of the upper atmosphere, ionosphere, and magnetosphere, is a key part of a scientific strategy that recognizes the importance of time series data relating to the geophysical environment. Optical, radar, geomagnetic, and other instruments provide an important view of external plasma processes, and data acquired simultaneously from many sites provide a basis for understanding many different manifestations of magnetospheric and ionospheric dynamics that are closely linked to the solar wind. Recommendations from the scientific community include increasing the number of operating stations, as well as modernizing them, to enable development of precise and high-quality databases.

Lack of observations from the Southern Hemisphere, particularly digital data, is a serious problem. These data are necessary to understand the asymmetries that arise from summer-winter differences in the polar ionospheres and from asymmetries in the geomagnetic field itself. Digital data acquisition in the Antarctic is particularly important, and the Antarctic is the only region where stable instrument platforms can be easily placed at very high polar latitudes (in the polar cap).

Equal emphasis must be given to global arrays and to dense regional arrays of instruments. Concern should be paid to including complementary instruments within arrays in order to achieve a rich source of fundamental parameters. Arrays must also be utilized to deconvolve the spatial and temporal aliasing of the data. This is a particular problem with the interpretation of data from a single station or spacecraft. Thus, the coordinated use of multiple stations and multiple instruments should become increasingly the norm in data analysis.

Laboratory Experiments

When a phenomenon has been identified by a spacecraft and the basic physics of it is not well understood, the laboratory is the ideal place to study it. The problems encountered in space observations of single-point measurements and nonrepeatability are overcome in the lab. A well-planned experiment can be carefully tailored so that it is repetitive in space and time. Plasma laboratory technology has advanced to the point that many experiments pertinent to space plasma phenomena can be performed. For example, in wave studies, waves can be made linear or nonlinear by the turn of an amplifier knob. Furthermore, these waves can be launched from one or more antennas and their fields mapped in the near and far zone. Beams can be introduced from localized sources, density nonuniformities can be repeatably produced, impurities can be added in known amounts at a given location, and plasma drifts can be created. Furthermore, measurements can be acquired at thousands of three-dimensional spatial positions and thousands of time steps during the interaction. This is impossible in space. Laboratory experiments can address both local and global physics issues (the latter are often determined by boundaries). In some cases, one can comprehensively analyze physical phenomena simultaneously from both global and local points of view. Furthermore, experimental devices may be rapidly configured to perform new experiments as ideas are developed. This can happen on the time scale of days or weeks, as contrasted with many years for satellites and several years for rockets. The hardware is reusable and flexible. Many different experiments can be performed on the same machine.

The challenge to laboratory plasma science is to continue to develop technology in order to extend the range of physical phenomena that can be studied. It is now possible to fabricate microscopic detectors and antennae that are capable of making spatially resolved, in situ measurements of the electric and magnetic fields in the plasma, the electron and ion temperatures, the plasma potential, and the velocity distribution functions. Nonperturbing optical techniques, such as laser-induced fluorescence and optical tagging, are now well established. Other new techniques are time-resolved tomography, electron cyclotron emission spectroscopy, and the use of the motional Stark effect. Three-dimensional probe systems can move detectors (optical or electronic) anywhere within large devices, so that full space-time data sets can be acquired. Visualization software and three-dimensional-graphics computers make analyzing these data possible.

Scientific areas in which laboratory simulation experiments can be carried out with current technology include properties of Alfvén waves, magnetic field line reconnection, wave-particle interactions leading to chaos, and current modulation of plasma conductivity.

CONCLUSIONS AND RECOMMENDATIONS

Space plasma physics, as the study of natural plasmas and associated technological applications, represents a vast multiscale physical domain with large variations in plasma sources, average thermal energy, flow velocities, magnetic field strength, and other underlying physical processes. As such, it represents an important regime for plasma science and technology and for our civilization. First, it provides us with an understanding in quantitative terms of the variety of interrelated complex processes acting to shape and influence our own terrestrial environment. Second, it affords the opportunity to observe at closer hand phenomena that may be operative in astrophysical situations. Third, space phenomena stimulate fundamental scientific questions relating to the behavior of plasmas under conditions that can be very different from those created and studied in terrestrial laboratories. And finally, space plasma science underlies the development of technological applications operating in or based on the space plasma environment. As a consequence, investigations of natural space plasma processes extend the frontiers of human knowledge, enabling broader physical understanding of plasmas within the context of their general behavior.

Progress to date in understanding the space plasma environment has provided us with a broad picture along with some detail. However, many important details of physical mechanisms remain unanswered, including interdependencies between sources and physical responses. Successful investigation of this environment requires a coordinated and balanced approach utilizing in situ observations, active experimentation, theoretical modeling, ground observations, and laboratory simulations. The requirement for a high degree of synergism is an inescapable conclusion.

The cornerstones of space plasma physics are observations carried out, analyzed, and interpreted in conjunction with complementary theory and modeling. Space plasma physics has historically developed in this mode. With the advent of new technologies, opportunities for further scientific understandings are nearly limitless. In studies similar to this one, the space community is currently examining future directions and independently identifying possibilities. The panel supports such efforts.

The use of space as a medium for active experimentation has declined to the point of near extinction. This is unfortunate, since active experiments may elucidate natural processes and expand in a unique way our basic understanding of the plasma state. The panel recommends a reinvigoration of the active experimentation area.

Meaningful laboratory experiments simulating space phenomena can now be performed in a number of different problem areas. Such experiments provide the opportunity to examine the relevant science in a controllable and reproducible manner; they are thus an important adjunct to highly transitory space observations and can hence serve as a vehicle for interpreting, substantiating, and/or planning the latter. Such laboratory experiments have been largely discredited in the past because they did not scale properly to space conditions, but that shortcoming has been circumvented by developments in technology. The panel recommends an initiative in laboratory experiments of sufficient magnitude to establish a small interactive community.

7

❖

Plasma Astrophysics

Plasma physics is relevant to almost every area of astrophysics, from magnetized, highly conducting stellar and interstellar plasma to gravitationally interacting many-body systems such as star clusters and galaxies. In some cases, the plasma physics is quite standard and requires only the application of known results. In other cases, the problem lies beyond the current frontiers of knowledge. Yet, plasma physics is not part of the standard graduate astrophysics curriculum, and plasma astrophysics has no distinct home at any federal funding agency. This chapter briefly describes some recent accomplishments and outstanding problems in plasma astrophysics, as well as education in and funding of plasma astrophysics.

RECENT ACCOMPLISHMENTS IN PLASMA ASTROPHYSICS

Any list of recent accomplishments is bound to be incomplete, but the work discussed below is representative.

Magnetized Disks, Winds, and Jets

Astrophysical interest in this problem goes back at least as far as the 1950s, when Fred Hoyle speculated that the early Sun could have transferred angular momentum to the protoplanetary disk via magnetic torques. The first quantitative theories began with the solar wind, which is observed to be magnetized. Simple models were developed to show that magnetic torques exerted by the solar wind could have removed significant quantities of angular momentum from the Sun and, by implication, that magnetized winds could play an important role in spinning down stars. Later, spurred by observations of accretion disks and jets around a wide variety of objects including protostars, white dwarfs, neutron stars, and black holes, astrophysicists developed models of magnetized winds and jets in disk geometry, included relativistic effects, strong magnetic fields, rapid rotation, and the effects of MHD waves and instabilities on the disks and the outflows. (See Figure 7.1.)

Particle Acceleration in Shocks

Although our understanding of high-Mach-number shocks is seriously incomplete, studies of particle acceleration in shocks have given us the best theories to date of cosmic-ray acceleration in the interstellar medium. The most notable successes of the theory are that it predicts approximately the correct power-law index of the energy spectrum, cosmic-ray intensity, and cosmic-ray composition (with the exception of the electron-to-ion ratio). Progress has been made on the analytical front through both kinetic and hydrodynamical descriptions of the particles and the shock and on the computational front through Monte Carlo simulations.

Magnetized Convection in Stars

The subject of stellar convection has a long history, since it was recognized many years ago that the radiative energy flux through a stellar envelope is limited by convective instability. Interest in the interaction of magnetic fields with convection stems from observations of the solar magnetic activity cycle and similar cycles on other stars, which show that magnetic fields are rapidly regenerated and reconfigured in the interiors of convective stars. Until recently, stellar convection was described only by dimensional arguments or mixing length theory. With the development of parallel and massively parallel computer architecture, it has become possible to simulate compressible convection in three dimensions and to include the effects of magnetic fields. Although the smallest relevant length scales are still unresolved by these calculations, the effects of buoyancy, concentration of flux into ropes, and dynamo activity—all processes that are believed to play an important role in the dynamics of stellar magnetic fields—are observed and can be studied.

Formation of Low-Mass Stars

It was recognized long ago that the ratio of magnetic flux to mass is much higher in the interstellar medium than it is in stars. It was proposed that interstellar clouds are supported against their gravitational fields by magnetic forces, that the fields slowly escape from the clouds by ion-neutral relative drift, and that the

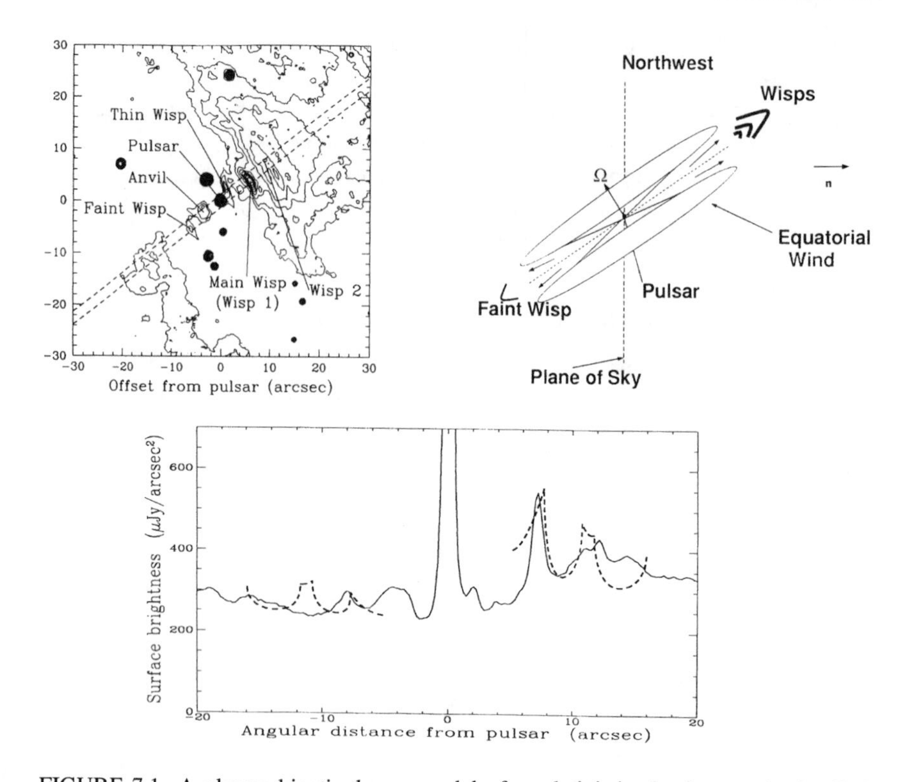

FIGURE 7.1 A plasma kinetic-theory model of a relativistic shock wave in the Crab Nebula. Upper left: Contour plot of the surface brightness of x-ray emission at 0.8 Å. The "wisp" features are thought to be visible manifestations of the otherwise radiationless outflow of rotational energy from the central pulsar. Upper right: Geometry of the outflow from the pulsar used in the construction of the theoretical model. The pulsar is assumed to lose energy in the form of a magnetohydrodynamic wind, flowing relativistically in an angular sector around the rotational equator of the pulsar. The magnetic field direction is orthogonal to the radial flow. The wind's composition is a mixture of electrons, positrons, and heavy ions, and it is quasi-neutral in the region upstream of the shock wave that terminates the outflow. Estimates indicate that a shock wave forms in the region of the observed wisps. The vector **n** points toward the observer. Lower panel: Comparison of the surface brightness (solid line) measured in the strip between the dashed lines in the upper panel with that predicted by the model (dashed line). The model represents the electron-positron pairs as a relativistically hot Maxwellian fluid, heated by the collisionless subshock at the leading edge of the shock structure. Heavy ions are modeled as a stream of particles gyrating in the electromagnetic field of the shock, compressing the magnetic field and pair plasma at each turning point of the ions' orbit. Each such compression appears as a surface brightness enhancement. The model successfully predicts the brightness of the faint wisp at –7 arc sec. (Reprinted, by permission, from M. Hoshino, J. Arons, Y.A. Gallant, and A.B. Langdon, Astrophysical Journal 390:454, 1992, and Y.A. Gallant and J. Arons, Astrophysical Journal 435:230, 1994. Copyright © 1992, 1994 by the American Astronomical Society.)

clouds collapse and form stars once a sufficient amount of their magnetic flux has been removed. This picture has been confirmed and extended by extensive theoretical calculations, including static models of magnetized, self-gravitating clouds; dynamical models of gravitational collapse; and both analytical and numerical calculations of ion-neutral drift.

PROBLEMS IN PLASMA ASTROPHYSICS

The panel has not attempted to make an exhaustive list of problems in plasma astrophysics, but instead has chosen a few problems that arise in a broad variety of physical environments, illustrating how plasma physics touches almost every part of astrophysics. Some of these problems are in the realm of space plasma physics as well.

Dense Stellar Plasmas

The mass density and temperature at the center of the Sun, which is an ordinary, low-mass star, are predicted to be about 100 g/cm^3 and 2×10^7 K, respectively. The energy produced by nuclear reactions diffuses outward as radiative energy, with most of the opacity due to bound-free transitions in elements heavier than helium. At these temperatures and densities, atoms are significantly perturbed by their nearest neighbors. Recent attempts to take these many-body effects into account when calculating the opacity and equation of state of dense stellar material have produced strikingly different results from earlier calculations, which has injected substantial uncertainty into models of solar-type stars and their evolution. This problem is at the intersection of plasma physics, statistical mechanics, and atomic physics.

Thermal Conduction in Plasmas

Observations suggest sharp temperature interfaces between the solar corona and lower atmosphere and at the boundaries of interstellar clouds. These interfaces are sharp in the sense that the inferred temperature scale height is comparable to the electron mean free path. The transport of heat becomes strongly nonlocal, and the electron distribution function becomes non-Maxwellian. Attempts to solve this problem have ranged from the application of theories of saturated heat flux regulated by ion-acoustic instabilities to attempts at full kinetic theory solutions of the Boltzmann equation.

Structure of Collisionless Shocks

Astrophysical shock waves are produced by energetic, impulsive events ranging from solar and stellar flares to sequential supernova explosions in associations of massive stars. Because mean free paths are long, these shocks must be collisionless. Remote sensing by spectroscopy shows that electrons as well as ions are heated to high temperatures. How is the ion distribution thermalized? How is this energy fed into the electrons? How is a small tail of particles accelerated to high energies, as is observed in the interplanetary medium? What is the back-reaction of the accelerated particles on the shock? These remain outstanding problems, because the Mach numbers are so high that the shocks are probably turbulent. Numerical simulations appear to be the most promising way to attack the problem at this point.

Acceleration of Particles to High Energies

Spiral galaxies appear to be permeated by a component of energetic particles, cosmic rays. In our galaxy the distribution function can be followed from subrelativistic energies to energies as high as 10^{21} eV. The most energetic particles cannot be confined by the galactic magnetic field. The energy density of these cosmic rays is similar to both the magnetic and the turbulent energy density in the galactic disk. How are these particles accelerated, and how do they propagate through the galaxy? The prevailing theories have particles at energies less than about 10^{15} eV accelerated by the Fermi mechanism in shocks and predict that they will be trapped within the galaxy by resonant scattering off Alfvén waves excited by their own anisotropy. For more energetic particles, the confinement is problematic, and the origin may be extragalactic.

Hydromagnetic Turbulence

There is abundant evidence for hydromagnetic turbulence in objects as diverse as stellar convection zones, the interstellar gas in galaxies, and the gas in clusters of galaxies. Turbulence can provide hydrodynamic forces (e.g., pressure support in interstellar clouds or acceleration in stellar winds), can lead to transport coefficients such as viscosity or resistivity that are much larger than their molecular values, and can provide significant heating through dissipation. Yet, we do not have a complete theory of hydromagnetic turbulence, and simulations, which are of great educational value, do not yet resolve the full range of relevant scales. Progress in understanding MHD turbulence will probably be made through a combination of direct observation (such as in situ measurements in the solar wind), simulations, and analytical theory.

Magnetic Reconnection

The magnetic Reynolds number (or Lundquist number) of astrophysical plasmas is typically huge, ranging from 10^8 in the solar interior to 10^{21} in the galactic interstellar medium. The naive conclusion is then that magnetic flux is permanently frozen into the plasma and that the field never changes topology. Yet, magnetic fields apparently do change topology (e.g., the star formation process seems to reconnect the magnetic field), and there is strong evidence that magnetic reconnection is an important source of energy in solar flares. How do field lines reconnect at very high magnetic Reynolds number? Present thinking suggests a two-stage process: some ideal magnetohydrodynamic effect creates steep gradients; then reconnection proceeds. We need a more fundamental understanding of the reconnection process itself; many of the fusion-oriented simulations have inappropriate boundary conditions for astrophysical systems. We also need a better understanding of the "ideal" current concentration phase.

The Magnetization of the Universe

Stars, galaxies, and the gas in clusters of galaxies possess magnetic fields. Standard cosmology predicts that the big bang did not produce a magnetic field. How and when did the universe become magnetized? Did large-scale, intergalactic fields form first and become incorporated into smaller structures, or did fields form first in stars, which then seeded their ambient medium through winds and supernova explosions? Are astrophysical magnetic fields nearly permanent, as suggested by their very long ohmic decay times, or are they constantly destroyed, regenerated, and reconfigured by turbulent dynamos?

Laboratory Experiments

There have been few laboratory experiments dedicated to plasma astrophysics, and any such experiments must carefully scale properly from the laboratory to the real astrophysical system. Areas in which experiments could be helpful include MHD turbulence, magnetic reconnection, shock waves, particle acceleration, dusty plasmas, and heat conduction. The status and future promise of laboratory experiments in many of these and related areas are discussed in Chapter 8.

TRAINING IN PLASMA ASTROPHYSICS

How do graduate students become equipped to deal with problems in plasma astrophysics? The standard graduate curriculum in astrophysics contains graduate physics courses, such as quantum mechanics, electrodynamics, statistical mechanics, classical mechanics—more or fewer, depending on the school and the inclination of the student. Then there are standard astrophysics courses, such as stellar structure and evolution, stellar atmospheres and radiative transfer, interstellar medium, and galaxies and cosmology. At many universities, no courses in plasma physics are taught in the physics or astrophysics departments. Such courses may be given in an engineering or applied science department, but these often have too technological an orientation to attract astrophysics students. Some plasma physics may or may not be integrated into one or more of the astrophysics courses, depending on the inclination of the instructor. Therefore, very few astrophysics students receive much formal exposure to plasma physics, and many astrophysicists view it as an arcane specialty.

Many astrophysicists would like to learn more plasma physics when motivated to do so by developments in their subject. For example, recent measurements of magnetic field strengths in dense, star-forming interstellar clouds have shown that the fields are large enough to strongly affect or even dominate the dynamics. This has spawned a real interest in MHD among interstellar medium researchers, and a number of people who ignored magnetic fields throughout most of their careers are now writing papers on them.

Such people would benefit from a good, modern text on plasma physics, one not oriented toward fusion plasmas, but stressing astrophysically interesting applications and using astrophysically relevant parameters and boundary conditions. Such a book could consist of chapters contributed by experts, provided that a good editor and refereeing system kept the quality high. Such a book could also be used for a graduate course or seminar.

FUNDING FOR PLASMA ASTROPHYSICS

Most plasma astrophysics by individual investigators is funded through the NSF and NASA. Some solar and space plasma physics has been funded by the Air Force and the Office of Naval Research, and some DOE funding has arrived through support for national centers.

Plasma astrophysics funding at the NSF suffers from a problem common to all of theoretical astrophysics: programs are organized by wavelength band or class or object, rather than by physical process. This discourages broad proposals. Yet, one of the exciting aspects of plasma astrophysics is that the same processes are at work under many different astrophysical conditions. Both the 1980 and 1990 NAS-sponsored decadal surveys of astrophysics (the Field[1] and Bahcall[2] Committees, respectively) recommended that the NSF establish a theoretical astrophysics program.

With the notable and important exceptions of its Astrophysical Theory and Space Physics Theory programs, NASA tends to support research focused on its missions. This has led to better support for space plasma physics, where much of the data is mission-relevant, than it has for plasma astrophysics, where fund-

[1]National Research Council, Astronomy Survey Committee, *Astronomy and Astrophysics for the 1980s*, National Academy Press, Washington, D.C., 1982.

[2]National Research Council, Astronomy and Astrophysics Survey Committee, *The Decade of Discovery in Astronomy and Astrophysics*, National Academy Press, Washington, D.C., 1991.

ing to develop new theories of distant objects is sometimes deemed an unacceptable stretch of the NASA mission. The panel would welcome an expansion of funding for basic research at NASA.

SUMMARY

Plasma astrophysics is an exciting and important area of astrophysics that is relevant to a wide variety of astrophysical phenomena and draws on an equally wide variety of topics in plasma physics. Problems in plasma astrophysics could stimulate important basic research. Yet the potential of plasma astrophysics is underrealized. The field is small and lacks the critical mass to provide graduate training at most institutions, especially given the lack of a suitable textbook. The funding base is diffuse, and plasma astrophysics is not recognized as a branch of astrophysics at any federal funding agency. Both the educational and the funding aspects need to be addressed to bring plasma astrophysics into the mainstream.

CONCLUSIONS AND RECOMMENDATIONS

Conclusions

Plasma astrophysics deals with phenomena and problems that are important to virtually every branch of astronomy and astrophysics. Some of these problems touch on areas that are central to basic plasma physics and have indeed inspired research in basic plasma physics. Yet, plasma astrophysics is not recognized as a coherent discipline by any federal funding agency.

Recommendation

The panel recommends that interdisciplinary programs be established at the National Aeronautics and Space Administration and the National Science Foundation with the goal of funding research in plasma astrophysics, whether through astronomy, solar-terrestrial research, physics, or computer sciences. The purpose of such programs would be to encourage research in plasma astrophysics, including research on basic processes that are relevant to many astrophysical systems.

PART III

❖

Broad Areas of Plasma Science

8

Basic Plasma Experiments

INTRODUCTION AND BACKGROUND

Plasma physics deals with the behavior of many-body systems under the influence of long-range Coulombic forces. Plasmas are inherently nonlinear media and, in the presence of a magnetic field, they are also anisotropic. Consequently, plasmas are capable of sustaining a wide variety of waves and instabilities. Plasmas can support three-dimensional currents and exhibit nonlocal behavior and "memory effects" (e.g., within the particle distribution functions). Plasma instabilities can lead to chaotic particle motions, to intricate wave dynamics, and to turbulence. Thus, understanding plasma phenomena involves fundamental aspects of statistical mechanics, fluid dynamics, electrodynamics, and frequently, atomic physics.

Progress in basic science has historically relied on a close interaction between experiment and theory. This is particularly true of plasma physics, where nonlinear and nonequilibrium phenomena in many-body systems are of central importance. In striking contrast to the central importance of laboratory experiments to this field, it is the finding of the panel that activity in and support for basic experiments has decreased markedly over the last two decades. For example, at the 1973 plasma physics division meeting of the American Physical Society, there were 126 papers on basic experimental plasma physics. In contrast, at the 1992 meeting, there was no general session on basic laboratory experiments, and there was only one poster session on laboratory experiments related to space plasmas. At this meeting, there were only about 30 experimental papers on basic plasma physics that were not related to a particular application, and half of these papers were on nonneutral plasmas. It is the conclusion of the panel that the level of activity in basic plasma experiment in the past 20 years almost certainly has been lower than it would have been if there had been in place a well-planned and balanced program in basic plasma science in the United States. The danger is that basic experimental plasma science will disappear in this country, unless one or more funding agencies assume the responsibility to support a critical mass of scientists in this area.

A survey of the plasma science community in the United States, conducted by the panel, shows that renewed support for basic laboratory plasma experiments is its highest priority. The panel has come to this same conclusion: The highest priority in establishing a healthy plasma science in the United States is renewed support for basic experimental research in plasma science. This conclusion coincides with the principal findings of the Brinkman report, *Physics Through the 1990s:* [1]

> Direct support for basic laboratory plasma-physics research has practically vanished in the United States. The number of fundamental investigations of plasma behavior in research centers is small, and only a handful of universities receive support for basic research in plasma physics. A striking example is the minimal support for basic research in laboratory plasmas by the National Science Foundation. . . . Because fundamental understanding of plasma properties precedes the discovery of new applications, and because basic plasma research can be expected to lead to exciting new discoveries, increased support for basic research in plasma physics is strongly recommended.

Support for basic plasma experimental research can be expected to serve an important educational function as well. University-scale experimental research programs in basic plasma science provide an excellent opportunity to train students in a variety of disciplines and techniques that are of importance in modern science and technology.

The chapters in Part II describe plasma physics experiments relevant to low-temperature and nonneutral plasmas, beams and radiation sources, and space and fusion plasmas. While many of these experimental studies have contributed significantly to our understanding of basic plasma science, they were often constrained by programmatic goals and by the plasma devices and plasma regimes relevant to a particular application. In this chapter, we focus specifically on what we have termed basic plasma experiments, whose primary goal is to isolate and study fundamental plasma phenomena in the simplest and most flexible situation possible. The objective of these experiments is to test our understanding of fundamental plasma phenomena, quantitatively and over the widest pos-

---

[1] National Research Council, *Plasmas and Fluids*, in the series *Physics Through the 1990s*, National Academy Press, Washington, D.C., 1986.

sible range of relevant plasma parameters. Although these experiments are not intended to focus directly on any particular application, they can be expected to provide a quantitative understanding of the underlying physical principles and to have a potentially significant impact on an entire spectrum of applications ranging from plasma processing and fusion to astrophysics.

Experiments on plasmas in the laboratory began in the 1830s with the work of Faraday to study the role of gas discharges in the chemical transformation of the elements. Further progress hinged on the discovery of the electron and the development of the atomic theory of matter at the end of the last century. In the 1920s, Irving Langmuir discovered the existence of collective oscillations in gas discharges. The understanding of plasma-related phenomena grew substantially with studies of electron beams in the 1940s and 1950s, in conjunction with the development of beam-type microwave devices. Since then, an enormous amount of work has been done in this area, and listing all of it is beyond the scope of this report. To convey the importance of a healthy and vital effort in basic experimental plasma science, we briefly review significant accomplishments in this area since 1980. We then proceed to discuss a number of important areas in which progress could be made in the next decade. These include topics that can be expected to have broad impact in virtually all of the areas of plasma science described elsewhere in this report. By the same token, basic experiments in specific topical areas are described in Part II. Examples include studies of electromagnetic wave-plasma interactions in the chapters on radiation sources and inertial confinement fusion and studies of fluid turbulence and transport in the chapter on nonneutral plasmas.

OVERVIEW OF RECENT PROGRESS

In this section, the panel presents a selection of areas and topics in which there has been significant progress recently both in experimental studies of fundamental plasma phenomena and in the development of new experimental capabilities.

Basic Plasma Experiments

Wave Phenomena

Bernstein Waves. Bernstein waves are predominantly electrostatic waves that propagate in a magnetized plasma. These waves require a kinetic description, since the dispersion relation is dominated by the cross-field motion of the plasma particles and the wavelengths of these waves are comparable to the gyroradii of the particles. There are branches of the Bernstein wave dispersion relation associated with each of the harmonics of both the electron and the ion cyclotron frequencies. Unlike acoustic and electromagnetic waves, Bernstein waves have no analogue in uncharged fluids, and they are therefore uniquely a plasma phenomenon. They are sensitive to kinetic effects and can be used as a diagnostic of plasma behavior as well as for plasma heating. An extensive body of knowledge has now been obtained from experiments that have used a variety of antennas and boundary conditions to elucidate the unusual properties of these modes. A wide variety of linear and nonlinear phenomena that involve Bernstein waves has been explored in the last decade, and they continue to be an important topic for basic research. Results from such studies have been used to interpret satellite observations of space plasmas. This knowledge has also been used to develop schemes for heating plasmas and for diagnosing plasma behavior. For example, there are potential applications using these waves to improve the stability and confinement properties of tokamak plasmas. However, an improved understanding of the nonlinear behavior of large-amplitude Bernstein waves will be required for such applications.

Mode Conversion. Understanding mode conversion has been an important area of investigation in the last decade. In finite-temperature, spatially nonuniform plasmas, there can be degeneracy in the wave dispersion near plasma resonances, and mode conversion can occur near the spatial locations of these resonances. In particular, long-wavelength waves, which are often electromagnetic in character, can convert into electrostatic waves that then convect away the wave energy. Consequently, mode conversion can provide an important physical mechanism for absorption of the energy of electromagnetic waves. A variety of cases have now been studied, including the conversion of electromagnetic waves to Bernstein waves, Langmuir waves, lower and upper hybrid waves, and whistler waves. However, several important issues remain to be addressed. For example, although the linear transfer of energy has been observed, quantitative studies of the converted waves, the efficiency of energy transfer, and the associated electric field patterns have yet to be done, and theories of these phenomena have yet to be tested quantitatively. Understanding mode conversion is of great practical importance because of potential applications to plasma heating and use in plasma diagnostics.

Wave-Particle Interactions

Magnetically Trapped Particle Instabilities. The ubiquitous spatial nonuniformities of magnetic fields in laboratory and naturally occurring plasmas can cause the generation of two distinct populations of plasma particles: passing particles and mirror-trapped particles. Under very general conditions, the bounce motion of the trapped particles can result in the spontaneous amplification of various plasma modes. Recent experiments, based on an arrangement of multiple mirrors, have now elucidated the fundamental nature of these processes.

Lower Hybrid Wave Current Drive. Lower hybrid wave current drive, a fundamental Landau-damping process, describes the transfer of the momentum of traveling waves, which have been excited by an external source, to the momenta of the individual plasma particles. By choosing an appropriate wave, it is possible to induce a dc current in the plasma by trapping particles in the wave. The first experiments were done in a linear device. Subsequent toroidal experiments have investigated this interaction in detail by exploiting the unusual properties of lower hybrid waves. Efficient methods of current drive will be important in developing a steady-state fusion reactor.

Beat Wave Excitation and Particle Acceleration. Basic laboratory experiments have demonstrated that when a plasma is irradiated by two electromagnetic waves whose frequency difference matches the local plasma frequency, very intense (GeV per centimeter) electric fields can be generated that travel at a significant fraction of the speed of light. Recently, it has been demonstrated in the laboratory that the controlled acceleration of a tenuous electron beam can result from its interaction with these plasma waves. Such studies suggest that compact particle accelerators based on this principle may be feasible. (See Figure 5.2.)

Nonlinear Phenomena

Double Layers. A fundamental nonlinear structure encountered in plasmas is the internal nonneutral sheath or double layer. A double layer can be thought of as the boundary between regions of plasmas having different particle distribution functions. An impressive body of experimental data has now been gathered from laboratory experiments on the shape, amplitude, and formation of these remarkable structures. These phenomena are important in space science. There are indications from satellite observations that double layers may form spontaneously in the near-earth plasma. The possible relationship between double layers and the formation of auroral beams is also being investigated.

Ponderomotive Forces and the Filamentation of Electromagnetic Radiation. The ponderomotive force is one of the basic nonlinear effects governing plasma behavior. This force can be thought of as arising from the added plasma pressure produced by the oscillatory motion of charged particles in a strong electromagnetic field. When the amplitude of this field varies as a function of position, the spatial variation in this additional contribution to the pressure results in the ponderomotive force. Several experiments have elucidated the macroscopic nature of the ponderomotive force, the limits of fluid-like response, and the limitations set by the requirements for adiabatic behavior. A variety of experiments in magnetized plasmas have explored how to use the ponderomotive force to quench various configurational instabilities and thereby to produce quieter and longer-lived plasmas with improved particle and energy confinement.

Experiments have been conducted to study the propagation of a high-power laser beam through a plasma and the resulting plasma response. These experiments have demonstrated the filamentation of the primary beam into high-intensity beamlets, which trigger secondary plasma-wave instabilities and create associated beams of fast electrons.

Magnetic Field Line Reconnection. The first magnetic field line reconnection experiments were done more than a decade ago in plasma pinch devices. Recently, a new generation of precise and well-controlled laboratory experiments has been carried out in which the ions are effectively unmagnetized but the electrons are magnetized. The magnetic field topology was mapped in three dimensions, and its dependence on plasma parameters was investigated. Observations include Alfvénic ion flow from the neutral sheet (i.e., a plane in the plasma at which the local magnetic field vanishes) and the formation of a neutral sheet on time scales less than the Alfvén transit time across the sheet. In the case where the current sheet was much narrower than its length, the breakup of the current sheet into a filamentary structure was observed. Other important nonlinear and three-dimensional phenomena were observed and studied, including the spontaneous generation of whistler-wave turbulence, the local formation of double layers, the generation of magnetic helicity, and the observation of highly non-Maxwellian particle distribution functions.

Recent experiments have also studied magnetic reconnection in the merging process that occurs when two spheromak plasmas are brought together. (See Figure 8.1.) These plasmas are isolated structures, spheroidal in shape, that are self-sustained by a combination of currents and magnetic fields. In this case, there are local current sheets with magnetized ions. These experiments indicate that the merging process depends qualitatively on the initial helicities (i.e., the "twists") of the magnetic fields of the plasmas involved in the merger process.

Plasma Reorganization. Several experiments have been done in the past five years on the merging of plasma currents and the propagation of currents across magnetic fields. (See Figure 8.2.) One common feature of these experiments is that the current flows are fully three-dimensional. For example, merging currents in a high-beta plasma (i.e., a plasma in which the plasma pressure is comparable to that provided by the confining magnetic field) were observed to spiral about each other as they coalesced. The currents evolved to become nearly parallel to the local magnetic field and hence force free. An elegant experiment in which an electron current was made to propagate across a magnetic field showed that whistler waves played a key role in the evolution of the current channel.

The experiments relied on highly reproducible, repetitive, plasma sources and on probes capable of studying the three-dimensional nature of the plasma behavior. These experiments are relevant to space plasma physics (such as the tethered shuttle experiment), solar physics, the study of helicity generation and helicity injection, and the behavior of three-dimensional current systems.

Chaos and Turbulence

Chaos. Accurate description of the plasma dielectric response relies on integration of the perturbation caused by an applied field along the trajectory of plasma particles. The perturbed currents and densities thus obtained may then be put into Maxwell's equations to determine the wave dispersion. However, even in a uniform magnetized plasma, the application of a single, finite-amplitude plane wave can be sufficient to render the particle orbits chaotic, and no self-consistent theory exists for the plasma dielectric response in this case. Experiments have now determined that non-self-consistent chaos theory correctly predicts several aspects of wave-induced particle chaos, as long as the wave amplitude is sufficiently small. Conservation laws describing the particle orbits, even during chaotic particle motion, have also been identified. Chaotic heating of plasmas has been observed, not only from externally launched waves but also from spontaneous, unstable waves in a plasma that is externally driven. These experiments were made possible by laser-induced fluorescence techniques that have advanced dramatically in the last decade.

Quasilinear Effects and Single-Wave Stochasticity. A series of experiments in single-component electron plasmas, which were carefully designed to eliminate the complications arising from ion dynamics, have tested the fundamental assumptions of "quasilinear theory," the standard model of weak plasma turbulence. These experiments demonstrated the importance of mode-coupling effects in modifying the wave-particle interactions described by the theory. In particular, in the presence of a mildly nonmonotonic particle distribution, unstable waves were found to grow and then saturate at the level predicted by the theory. However, the growth rates of individual waves were found to depend on the rates at which other waves grew, and this is *not* accounted for in the theory. Thus, a complete understanding of this important problem has yet to be achieved. This topic is related to the common assumption of the "random phase approximation" in turbulence theory, which is central to current descriptions of weak turbulence. The potential for new experiments in this area is discussed below in the context of turbulence and turbulent transport.

Complementary experiments have observed the evolution of a large-amplitude monochromatic wave to a stochastic signal, via sideband generation and trapped-particle dynamics. A very important, but as yet unresolved, question is the detailed mechanism by which a single, large-amplitude wave is transformed into the background of weak turbulence that can be addressed by quasilinear theory.

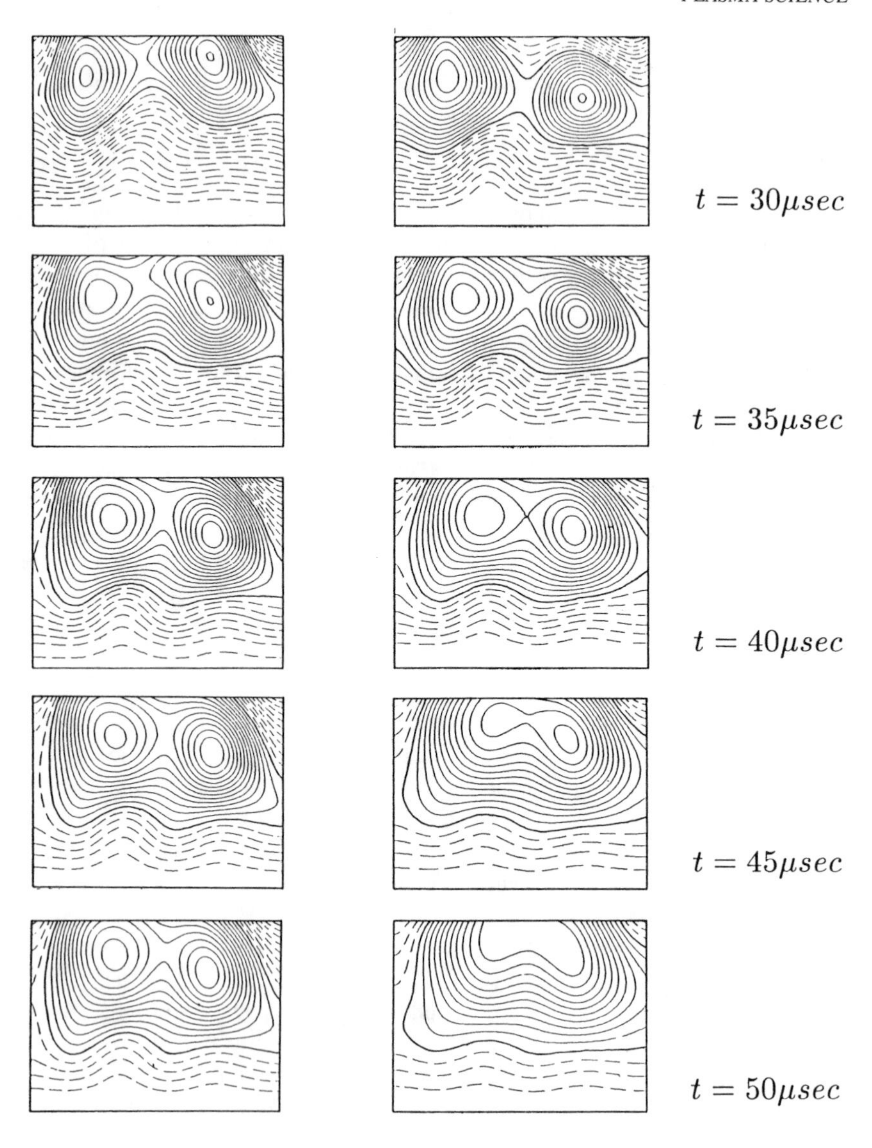

FIGURE 8.1 Experimental study of magnetic reconnection processes in the merging of two spheromak plasmas. This experiment demonstrated that the three-dimensional structure of the magnetic field is crucial to the merger process in that the difference between the co-helicity and counter-helicity merger process is due to the relative directions of the out-of-plane components of the magnetic field in the two plasmas. A new mechanism of

Collisionless Heat Transport. Laboratory experiments have now explored the important question of how heat is transported in collisionless plasmas. These measurements involve the application of high-power microwave beams to generate hot electron tails in a nonuniform plasma. The qualitative features of this effect and the important scaling properties have been identified. They have helped to clarify the relevant theoretical issues in this area.

Strong Langmuir Turbulence. One of the significant advances in the understanding of nonlinear plasma behavior has been the development of the concept of plasma-wave collapse and the associated spiky turbulence that frequently accompanies it. Several laboratory experiments, aimed at uncovering the microscopic dynamics of Langmuir-wave collapse, have used both electromagnetic driving and electron beams to trigger the collapse of extended wave packets, which in turn produces strongly localized fields and density depletions or cavitons. More recently, the ionosphere has been used to demonstrate the ubiquitous nature of these phenomena and the important role they play when a plasma is driven by large-amplitude perturbations.

Experimental Techniques and Capabilities

Opportunities for advances in experimental physics are often linked to the development of new technologies. The effect of these technologies is twofold. First, they enable the creation of experimental conditions that permit the demonstration and isolation of important physical effects. In plasma science, this frequently involves both new means of plasma production and new means of plasma confinement. In addition, new technologies frequently lead to new diagnostic techniques and new means of processing data, which not only result in improved accuracy and precision but often result in new perspectives on the underlying physics.

Plasma Sources

Over the past 20 years, there has been substantial progress in the development of improved, quiescent plasma sources. Some of the first, "high-quality" plasmas used in basic research were created in Q machines (Q stands for "quies-

---

plasma acceleration (perpendicular to the plane of the figure) was discovered in the course of this work. (Reprinted, by permission, from M. Yamada, F.W. Perkins, A.K. MacAulay, Y. Ono, and M. Katsurai, Physics of Fluids B 3:2379, 1991. Copyright © 1991 by the American Institute of Physics.)

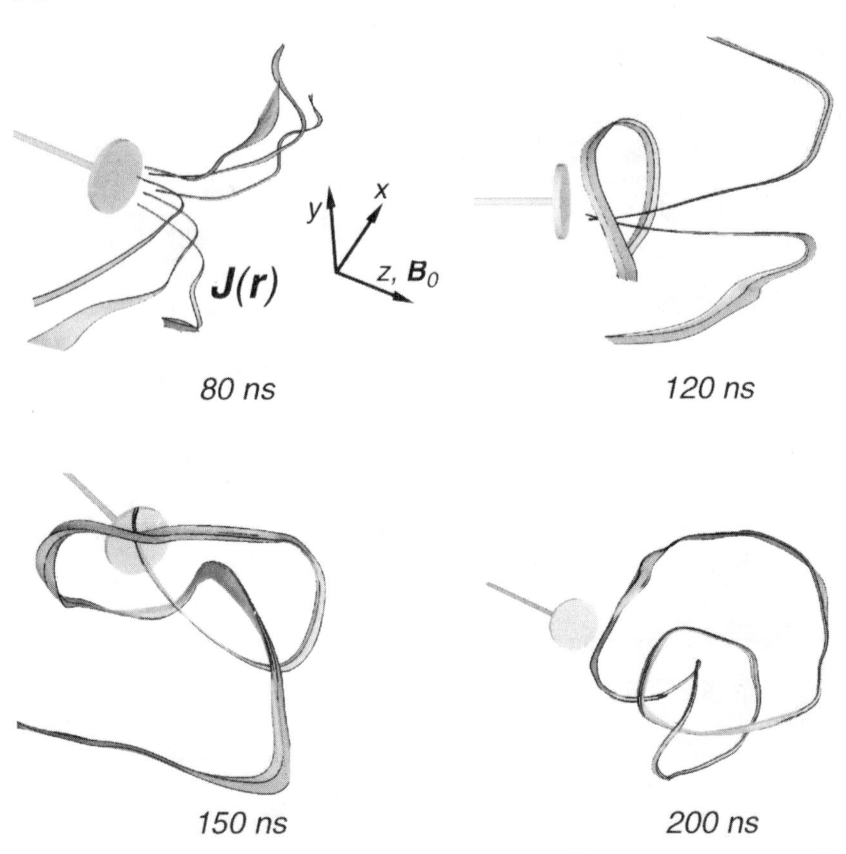

FIGURE 8.2 Experimental study of the penetration of a pulsed current into a magnetized plasma. Shown are the characteristic field lines, sheets, and tubes of the current density, $J(r)$, at different times after a 100-ns (FWHM) current pulse is applied to a disk electrode (shown). These data are extracted from a dataset of 10,000 point measurements at each time step. Typical experiments involve studying 1000 such time steps. At 80 ns, the current penetrates a short distance from the positive electrode into the plasma, before turning back to the negative electrode located at the back endwall of the vacuum chamber. Little helicity is observed in this fountain-like current flow. As the current propagates (120 ns) two distinct current systems are observable: a closed azimuthal Hall current in regions where $J_z \approx 0$ and field-aligned solenoidal plasma currents between the positive and negative electrodes. At 150 ns a current tube starts off-axis, where $J_z \neq 0$ and $J_B \neq 0$, and exhibits strong helicity; i.e., it twists and knots in the right-hand direction. After the end of the applied current pulse, at 200 ns, the current lines detach and propagate away from the electrodes, and shown is a closed, singly-knotted, twisted current tube. Experiments like this, which illustrate the fully three-dimensional nature of the dynamics of the plasma response resulting from such a current pulse, have recently been made possible by the advent of fast, relatively inexpensive laboratory computers with large data handling capabilities. (Courtesy of R. Stenzel and M. Urrutia, University of California, Los Angeles.)

cent"), which were developed 30 years ago. These devices generate a magnetized plasma column, with a diameter of about 10-20 ion Larmor radii, that is well suited for the study of such phenomena as drift waves and ion cyclotron modes.

The plasmas in Q machines are such that the electrons and ions have equal temperatures (i.e., $T_e = T_i$). Consequently, these devices are not appropriate for the study of ion acoustic waves, which are strongly damped in such plasmas. The use of large numbers of small, permanent magnets to produce surface magnetic confinement, together with a variety of different electron sources, has provided a way to produce unmagnetized, collisionless plasmas that are both isotropic and quiescent. Such devices have $T_e/T_i \approx 10$, and they are well suited to the study of the linear and nonlinear behavior of ion acoustic waves. These plasma devices have also permitted experiments on plasma sheaths and on a variety of other linear and nonlinear waves. Combination of two or three of these plasmas has resulted in so-called double and triple plasma devices that have been used to study beam-plasma interactions, solitons, and electrostatic shocks.

In the past decade, dc discharges based on oxide-coated cathodes have resulted in the ability to produce large, quiescent, magnetized plasma columns, of the order of 50 cm in diameter (which is equivalent to 500 ion Larmor radii) and 10 m in length. Efficient, microwave-generated plasmas are now also conveniently available. Electron cyclotron resonance sources provide another way to study highly collisional plasma phenomena, with ion-neutral mean free paths of several centimeters or less. Inductive sources have recently shown considerable promise in producing uniform, unmagnetized and magnetized plasmas in the pressure range greater than 5 mtorr and, for example, have already been employed in studies of double layers.

During the last few years, "helicon" sources (bounded whistler-wave sources) have produced steady-state plasmas with densities as high as 10^{14} cm^{-3}. Such sources, which operate between the lower hybrid and the electron cyclotron frequency, do not have a high-density cutoff; they are therefore useful in producing plasmas with high densities.

Plasmas consisting of negative and positive ions, with very low concentrations of electrons, have also been created, both with and without a magnetic field. The production of these plasmas relies on the large electron-attachment coefficient of gases such as SF_6 for cold electrons. For sufficiently low values of the electron density, waves and instabilities in these plasmas can differ qualitatively from those in electron-ion plasmas, since the dominant charge species now have comparable masses.

Mechanical Probes

Refinement of probe techniques has occurred hand in hand with plasma source development. These include directional velocity analyzers (with resolutions of a few degrees in velocity space), emissive probes, and electric dipole probes that are sensitive to the total electric field, including the magnetic component (i.e., $\boldsymbol{E} = -\nabla\Phi - \partial\boldsymbol{A}/\partial t$). Computer-controlled probe drives capable of full, three-dimensional motion have been developed, and essentially the first-ever fully three-dimensional studies of several important plasma phenomena have been conducted.

Laser-Based Optical Diagnostics

The potential of laser-based optical diagnostics for plasma science has continued to develop. Two general categories of recent achievements are measurements of the spectrum of collective plasma density fluctuations and measurement of the single-particle distribution function, both by Thomson scattering and by laser-induced fluorescence. Scattering from collective plasma fluctuations has been used to study both thermal and nonthermal effects, such as waves and instabilities and entropy fluctuations. A variety of imaging diagnostics, such as phase-contrast and shearing-plate interferometry, has been developed for use in experimental plasma research. Single-particle scattering techniques have been developed to study both the ion and the electron velocity distributions. Thomson scattering techniques have been developed to measure electron density and temperature with unprecedented spatial and temporal resolution.

The use of laser-induced fluorescence techniques to study plasma ions has been one of the major recent developments in plasma diagnostics. Not only does this technique permit temporally and spatially resolved measurement of the ion distribution function with high sensitivity, but it also permits a variety of extensions. (See Figure 8.3.) For example, the use of metastable states provides the capability of measuring nonlocal plasma properties, and metastable or spin-polarized ions produced by optical pumping may be used as test particles to trace ion orbits and to study particle transport.

Laser-induced fluorescence has now been used to measure ion distribution functions in a variety of physical processes and to study the plasma dielectric response, and optically tagged test particles have been used to measure Fokker-Planck coefficients for collisional diffusion. Such test particles have also been used to measure the Lyapunov exponents that characterize chaotic particle motion and to measure the transport arising from a variety of physical processes.

Recent developments in laser and optical technologies include the development of new, high-performance laser materials, such as Ti-sapphire, and a wide variety of solid-state lasers. Other important developments include nonlinear optical materials, which are essential to the production of tunable radiation at short wavelengths via frequency multiplication. Improvements in short-pulse laser technology have increased the time resolution of measurements to better than 10^{-14} s, which is shorter than virtually all of the natural time scales in most laboratory plasmas.

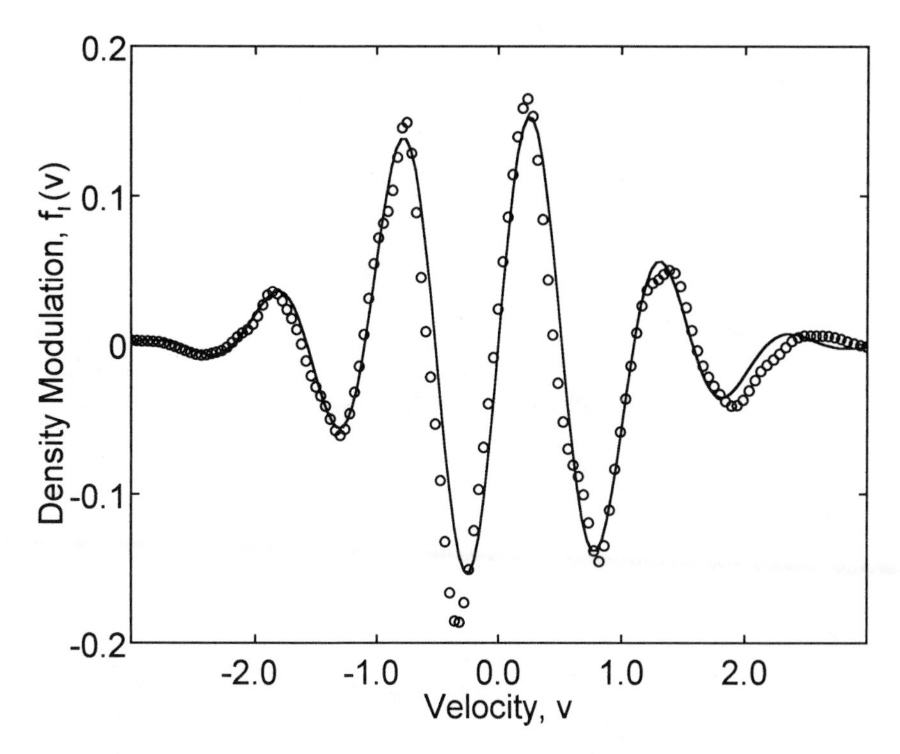

FIGURE 8.3 The oscillatory part of the ion velocity distribution, associated with a wave in a magnetized plasma, is shown as a function of the velocity of the ions. The data (open symbols) were measured nonperturbatively at a point in the plasma with recently developed laser-induced fluorescence techniques. For comparison, the solid curve shows the theoretical prediction, assuming an ideal plane wave. Experimental techniques such as the one illustrated are now capable of directly studying detailed aspects of plasma behavior, including plasma flows, chaotic particle motions in response to large-amplitude waves, and particle transport due to waves and turbulence. (Courtesy of F. Skiff, University of Maryland.)

Data Acquisition and Processing

Improvements in data acquisition systems and probe drives now allow the recording of fully three-dimensional data sets with good spatial and temporal resolution. Such data are now routinely displayed with relatively inexpensive, powerful workstations. The current state of the technology permits the exploration of many important physical processes that are of fundamental interest to basic plasma physics and are relevant to a variety of applications.

RESEARCH OPPORTUNITIES

Fundamental Plasma Processes

The following is a selection of important fundamental problems in basic plasma science that could be addressed in the next decade by a new generation of plasma experiments.

Wave Phenomena

Alfvén Waves. Alfvén waves are modes of oscillation of a magnetized plasma at frequencies below the ion cyclotron frequency. They are important both in fusion plasmas and in space plasmas, such as the solar wind and the Earth's magnetosphere and ionosphere. Alfvén waves can act to transport information about magnetic field disturbances in magnetized plasmas. Alfvén waves and their high-frequency analogue, magnetosonic waves, are important in magnetic confinement fusion research, where they are candidates for plasma heating and noninductive current drive.

In spite of their importance, relatively little work on these waves has been done in the laboratory because Alfvén waves have relatively long wavelengths (1-5 m) in plasmas of reasonable density. Recently, well-diagnosed plasmas have been developed that are sufficiently dense and large enough to accommodate several Alfvén wavelengths. Thus, carefully controlled laboratory Alfvén wave experiments are now possible. Topics currently under investigation include the dispersion relation for these waves, their reflection properties and spatial structure, and the nonlinear behavior of these modes.

Wave-Plasma Interactions. Wave-plasma interactions have been under active investigation, but many important questions remain. Outstanding issues include the modulation of plasma currents by waves and the modulation of low-frequency waves (e.g., Alfvén and whistler waves) by local fluctuations in the plasma conductivity. Another important problem that has not yet been studied experimentally is resonant absorption in the situation where a wave propagates along a density gradient that is parallel to the magnetic field. Important effects include nonlinear refraction and the generation of non-Maxwellian electron distribution functions and Langmuir turbulence. Such experiments are relevant to fusion physics as well as to heating of the F-region of the ionosphere. Information generated by these experiments also may be relevant to the development of advanced particle accelerators and to the generation of intense electromagnetic waves.

Intense Laser-Plasma Interactions. As described in Chapter 5 on beams and radiation sources in Part II, recent technological developments have led to the development of compact, subpicosecond terawatt lasers with beams that can be focused to intensities greater than 10^{18} W/cm^2. At these intensities, an electron is accelerated to relativistic energies in one period of the laser light. This permits the study of highly nonlinear, fast, relativistic processes in laser-plasma interactions. If the laser is focused on an overdense plasma target, dc magnetic fields of the order of 10^9 G are predicted to occur. It will be important to determine whether the nonlinear ponderomotive forces and relativistic effects can reduce the diffraction of these ultrahigh-intensity laser beams, so that the light can be focused to beam sizes smaller than a few Rayleigh lengths, the limit expected at lower values of light intensity and in a linear medium.

Chaos, Turbulence, and Localized Structures

Nonlinear Particle Dynamics and Chaos. Modern concepts of nonlinear dynamics have created a renaissance in classical physics, bringing new techniques to bear on long-standing problems. One crucial issue in plasma physics is the onset of chaotic particle motion in response to coherent or turbulent wave fields. Of particular interest are a self-consistent description of the system under such circumstances and the evolution of the system from regular particle motion to chaos. It is now possible to conduct precisely controlled experiments in the laboratory to address these important problems, which are of interest in a wide variety of contexts, ranging from fluid dynamics to advanced particle accelerators.

Nonlinear Wave Phenomena. With the exception of Alfvén waves, most of the other linear branches of the plasma dispersion relation have been explored. In addition, many nonlinear, three-wave coupling processes have been observed. However, the transition from linear to turbulent wave behavior is not understood. This includes the nonlinear behavior associated with almost every branch of the plasma dispersion relation.

Turbulence. Very generally, plasmas are electrodynamic, many-body systems far from equilibrium that are dominated by nonlinear effects. Consequently, plasmas are typically highly turbulent, exhibiting large fluctuations in such quantities as the local density, temperature, and magnetic field, which can vary rapidly in time and space. Important examples of plasmas whose behavior is influenced profoundly by turbulence include essentially all magnetically confined fusion plasmas and many astrophysical and space plasmas. We have no first-principles understanding of turbulence in *any* plasma, and understanding such turbulent behavior is perhaps the key unsolved problem in plasma physics. This problem presents an important synergism with fluid dynamics, in that plasmas can often be modeled as fluids and understanding turbulence is central to a complete description of fluid systems.

Since turbulence is so common in plasma physics, the potential rewards for achieving predictability are particularly high. In the past decade, new plasma sources and measurement techniques have been developed that will allow us to undertake a new generation of precisely controlled experiments to study turbulent plasma behavior. One starting point in achieving a deeper understanding of turbulence will be further study of the questions raised by the observed breakdown of quasilinear theory and experimental tests to determine the range of validity of the random phase approximation.

Turbulent Transport. Profound consequences of plasma turbulence include the transport of both particles and energy and the acceleration of particles that can be induced by turbulence. Such transport can completely dominate plasma behavior. For example, transport by turbulence, in the form of both convection and enhanced diffusion, is the dominant transport mechanism of particles and energy in present-day tokamak fusion plasmas. Turbulent transport presents an excellent opportunity for carefully controlled laboratory experiments. At least in low-temperature laboratory plasmas, techniques are now available to study both the transport of particles and energy and the fluctuations responsible for this transport. To establish the causal connection between turbulence and transport, it will be necessary to make precise, spatially resolved measurements of fluctuating quantities such as plasma temperature, density, velocity, and magnetic field and to establish the correlations between these quantities and local measurements of the particle and energy fluxes.

Because turbulence and turbulent transport are not understood in any plasma, careful experimentation in flexible, small experiments is likely to make significant contributions to testing existing theoretical predictions and to guide further theoretical work in this important area. Given the fundamental lack of understanding and the important practical consequences that would derive from a deeper understanding of turbulence and turbulent transport, a sustained program of both theoretical and experimental research is extremely important.

Sheaths, Boundary Layers, and Double Layers. Plasma sheaths (i.e., regions where the plasma is not charge-neutral) have been an important topic throughout the history of plasma physics. All plasmas in the laboratory and in space have boundaries at which there are sheaths, and probes and antennas immersed in plasmas are surrounded by such sheaths. One important area for future research is the nature of sheaths in magnetized plasmas. To probe the structure of such sheaths requires detectors smaller than an electron Debye length. Such probes and probe arrays can be expected to be available in the next few years.

Double layers are a class of sheaths that are detached from a physical boundary and are supported by locally non-Maxwellian conditions. Although there has been some research done on double layers, there has been little work on situations in which the ions are magnetized and situations involving the transition regime between collisionless and collisional plasmas. This is of importance for solar physics (in regard to coronal holes) and in ionospheric heating experiments. Double layers are responsible for localized particle acceleration and could also play an important role in the aurora.

Shock Waves. Much laboratory work has been done on shocks in unmagnetized plasmas. Shocks also have been studied in pinches and exploding wires. However, careful experiments on Alfvénic shock waves in magnetized plasmas have yet to be done. Of particular interest is the propagation of large-amplitude (i.e., $\delta B \approx B$) magnetic pulses. Work would include studies of wave steepening, particle reflection and heating, and a search for a new class of shocks (the "intermediate shock") that has been predicted but not yet observed. Such shock wave phenomena are of importance in space and astrophysical plasmas.

Striated Plasmas. Plasmas with nonuniformities, such as density or temperature striations in the direction parallel to the magnetic field, are of fundamental interest. They occur, for example, in the ionosphere and in the aurora. If the gradient in the plasma properties is steep compared to the wavelengths of interest, these structures can trigger the mode conversion of whistler waves. The reflection, refraction, and interaction of waves with plasma structures that have steep gradients has not been studied in the laboratory and presents a difficult "plasma scattering" problem. Striated plasmas are not limited to those with density perturbations but also include local "hot spots" and magnetic field perturbations. Topics of interest include the interaction with lower hybrid waves, refraction and reflection, fast-particle generation, and minority-species heating.

Flows in Magnetized Plasmas. It now is possible to generate highly magnetized laboratory plasmas in which the diameter of the plasma column is much larger than the ion Larmor radius (e.g., by factors of as much as 10^3) and in which magnetic Reynolds numbers of 10^4 to 10^5 are attainable. (The magnetic Reynolds number is the time scale for transport of the magnetic field by the flowing plasma relative to the time scale for diffusion of the field due to the finite resistivity of the plasma.) By using multiple sources (so called "double-plasma" configurations), flowing plasmas can be generated with drift velocities comparable to the Alfvén wave velocity and with Mach numbers (i.e., plasma flow velocities relative to the ion sound speed) of the order of 500. Large currents can be entrained in these plasmas. Such situations are predicted to lead to shocks and turbulence. Insights into dynamo action (discussed below) are also likely to be achieved in such experiments. These experiments also would be relevant to the solar wind and to other solar and astrophysical processes.

Plasmoids. "Plasmoids" are plasma structures that propagate as recognizable entities through a background plasma. Satellite data suggest that plasmoids may occur in the Earth's magnetotail and become detached and move away from the Sun during magnetic storms. The propagation of plasmoids has been studied with small plasma guns. Larger structures of interest to fusion physics (i.e., spheromak plasmas) have also been investigated. (See Figure 8.1.) The latest generation of diagnostics, coupled with the recently developed ability to generate plasmoids easily, now permits a new generation of experiments. For example, one now can study in detail how plasmoids are generated, how they propagate, and the details of their internal field structure and associated plasma currents.

Magnetic Effects

Magnetic Field Line Reconnection. Magnetic field line reconnection is one of the principal means by which magnetic field energy is converted to thermal energy and plasma motion. For example, it is thought to be responsible for the high temperature of the solar corona and to be important in the Earth's magnetotail and in many astrophysical situations. Magnetic reconnection also is of importance in understanding the so-called sawtooth crashes that occur in the hot core of tokamak plasmas when a certain type of magnetohydrodynamic instability is present. In this case, reconnection has the effect of expelling hot plasma from near the plasma center and is detrimental to plasma confinement.

Experiments on magnetic reconnection in plasmas with unmagnetized ions and magnetized electrons have already been done. Areas for further study include cases where the magnetic Reynolds number is greater than 100. Important issues include the three-dimensional nature of this phenomenon, the connection between global and local time scales, the acceleration and heating of the plasma particles, and the generation of plasma flows.

Dynamo Action. The dynamo is a process by which the kinetic energy of a conducting fluid is transformed into magnetic field energy. (See Figure 8.4.) In a dynamo, a "seed" magnetic field from a small current fluctuation can be stretched and reconnected by the turbulent fluid motion. In principle, this can lead to amplification of the magnetic field to a level where the magnetic field dominates the dynamics of the fluid flow. The dynamo is a fundamental process in magnetohydrodynamics, and dynamo action is crucial to understanding many aspects of space physics and astrophysics. For example, it is believed to be the origin of the magnetic fields of such diverse objects as the Sun and the accretion disks of stars and is intimately connected with the physics of novae and supernovae.

The conditions for dynamo action require large-scale flows in highly conducting media, and up until now, such conditions have proven difficult to achieve in the laboratory. The criterion for dynamo action is the achievement of Reynolds numbers on the order of 100. Possibilities now exist to carry out well controlled

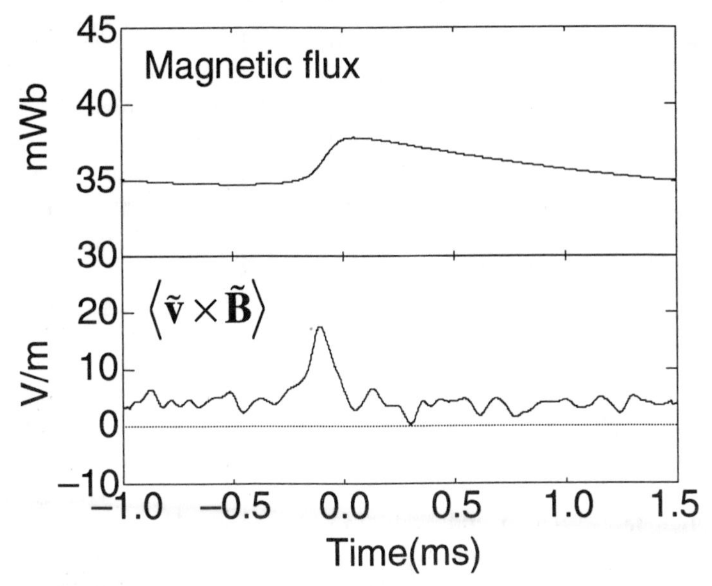

FIGURE 8.4 Demonstration of the dynamo effect in a laboratory plasma. The dynamo effect is the spontaneous self-generation of magnetic fields within plasmas, a process common in astrophysical bodies such as the Sun. Shown is a "dynamo event" in which magnetic flux is suddenly generated in a toroidal laboratory plasma. By measuring the local electromotive force, $\langle \mathbf{v} \times \boldsymbol{B} \rangle / c$, generated by fluctuations in plasma velocity and magnetic field, it was established that this force is responsible for generation of the magnetic flux. The fluctuating flow and field were measured with Langmuir and magnetic probes inserted into the plasma edge. This particular dynamo mechanism has long been thought to be an important source of astrophysical magnetic fields. (Reprinted, by permission, from H. Ji, A.F. Almagri, S.C. Prager, and J.S. Sarff, Physical Review Letters 73:668, 1994. Copyright © 1994 by the American Physical Society.)

dynamo experiments, for example by establishing a rapid fluid flow in a large vat of liquid sodium. Such experiments can provide new and fundamental insights into the nature of dynamo action and provide a quantitative basis for refined theories of many important physical processes.

Magnetic Reconfiguration. Many plasma instabilities and processes are inherently three-dimensional. Laboratory experiments now allow these processes to be explored in detail. Important topics include the dynamics of three-dimensional current systems, current and wave filamentation, current sheet formation, the effect of magnetic forces on plasma currents, helicity generation, and helicity conservation. These processes have far-reaching consequences in the understanding of solar flares, solar magnetospheric physics, and fusion physics.

New Experimental Capabilities

In any experimental science, and particularly in physics, advances in diagnostics consistently have led to new discoveries and frequently have opened up entirely new areas of research. Several new tools for plasma research are now becoming available. Some of these techniques have not been developed with plasma diagnostics in mind, but they can be expected to have significant impact on experimental plasma science. In many cases, progress is likely to require the collaboration of plasma physicists, solid-state physicists, and engineers. Much of this work will have important applications in other fields.

Use of Nanotechnology

Advances in nanotechnology are likely to have a profound impact on experimental plasma physics. Typical devices are miniature valves and mass analyzers. Techniques widely used in the semiconductor industry will enable the production of particle detectors, mass-sensitive energy analyzers, and magnetic and electric field probes with overall scale sizes smaller than 1 mm and active sensor areas less than 10 mm in diameter. These detectors will be capable of providing spatially resolved measurements on the scale of the Debye length and electron cyclotron radius in research plasmas with densities of the order of 10^{12} cm^{-3}, electron temperatures of tens of electron volts and magnetic fields of 0.1 T. Such probes would produce a minimal perturbation of the plasma if their connections and supports were also microscopic (≈ 0.2 mm). It is likely to be possible to position many (e.g., 10^4 to 10^6) of these detectors on a lattice that could be moved within the plasma.

Optical Diagnostics

The recent discovery of giant Faraday rotation in magnetoactive crystals now enables the construction of magnetic field probes as small as 10 μm in diameter and less than 1 mm long. As a light beam traverses one of these small crystals, the plane of polarization of the light is rotated. Preliminary tests have demonstrated sensitivities of 1 G per degree of angular rotation and response times faster than 10^{-9} s. Such probes are immune to electrical pickup and are nonmetallic. Another important capability has been created by the discovery of the quantum-well effect in crystals. Quantum-well devices can now be fabricated into microscopic probes to measure the local amplitude of the electric field. Arrays of these optical probes could be used to diagnose the space-time behavior of the electric fields associated with plasma waves and currents.

New Regimes of Plasma Parameters

As described in Part II, advances in laser technology now make possible laboratory experiments in previously inaccessible regimes of plasma parameters. Both short-pulse, high-power lasers and multiphoton ionization using tuned sources can be used to produce liquid or solid density plasmas, in which both quantum and classical many-body effects are important. The creation of these high-energy-density plasmas also opens up the possibility of studying plasmas with highly ionized ions (i.e., high-Z plasmas).

Data Acquisition

In the past 10 years, experimental physics has benefited greatly from advances in digital technology. Analog-to-digital converters and microprocessors have decreased drastically in price. Workstations are now available with 128 Mbyte of memory and 8 Gbyte of disk storage, and this trend shows no sign of saturating. A system with 10^6 channels of acquisition is capable of acquiring on the order of 1 Gbyte of data per second. Such a data acquisition system might consist of many parallel processors sharing a fast network and have 10 Gbyte of random access memory and 10 to 100 Tbyte of mass storage. This system would permit the study of nonuniform and fully three-dimensional plasma phenomena and plasma processes occurring on more than one spatial scale with unprecedented spatial and temporal resolution.

Massively Parallel Plasma Diagnostics

Interactions among plasma particles range from short-range collisions between individual particles to long-range, collective forces; consequently, plasmas frequently contain several different characteristic length scales. Magnetized plasmas are inherently anisotropic and nonlinear, exhibiting nonlocal behavior, chaotic particle motions, turbulence, and self-organization. The fundamental equation describing the plasma behavior of a many-body system of N charged particles is Liouville's equation for the distribution function in the $6N$-dimensional phase space of the system. As a practical matter, theoretical descriptions of plasmas are frequently based on much simpler and more tractable equations. However, the assumptions concerning the statistical structure of the plasma, which buttress the derivations of these simpler treatments, have not been tested.

In the next decade, a new generation of plasma experiments is likely to be able to make significant contributions in testing the validity of the approximations used to describe plasmas, for example, as fluids or as many-body systems described by kinetic theory or by a particular kind of particle correlation function. With what is now or will shortly become available, experimental plasma science will be able to explore a range of plasma problems with a precision and to a degree of detail previously unattainable. Below, we briefly list some possibilities, assuming the capability exists to create a lattice of thousands of microscopic detectors and/or thousands of channels of optical probes.

In plasma physics, the plasma is typically described by a combination of time-averaged and fluctuating fields. In the case of a turbulent plasma, the average of a quantity may be much smaller than its fluctuating component. If one considers an experiment that is repeated many times, all individual quantities, such as the instantaneous values of the electric and magnetic fields, the density, and the particle distribution function, can in principle be measured and recorded. This detailed set of measurements could be used to calculate the higher moments of the distribution function to test the assumptions that go into the derivations of the equations of kinetic theory. For example, our present understanding of three-body correlations is poor, but such quantities could be measured directly.

Fine structure in the particle distribution functions could also be measured. Measurements by particle detectors on spacecraft and in the laboratory indicate that when instabilities are present, the distribution functions cannot be regarded simply as functions of the magnitudes of the components of velocity perpendicular and parallel to the magnetic field. Snapshots of the distribution functions could be expected to reveal phase-space structures that go far beyond such a simplified description. Such highly anisotropic particle distribution functions can be expected to have profound effects on the growth and damping of a variety of plasma waves. With such detailed measurements, one could also test the validity of equating temporal averages with spatial ones.

The ability to probe fine spatial scales will permit a detailed exploration of plasma sheaths and boundary layers. For example, tiny puff valves and microbeam sources could be used to tailor the local particle distribution function or to add minute quantities of an impurity ion. Finally, phenomena on scale lengths ranging from less than the Debye length, to the ion cyclotron radius, to the electron cyclotron radius, could be explored simultaneously in one experiment. This would allow exploration of physics from the regime in which single-particle interactions are important to the regime in which kinetic and MHD effects are dominant.

SUMMARY, CONCLUSIONS, AND RECOMMENDATIONS

The panel has great concern that basic experimental plasma science is disappearing in the United States. By its count, there are currently fewer than 20 groups engaged in basic plasma experiments in the United States. Yet an intellectual atmosphere that allows for dialogue, complementary experiments, and in some cases, competition is necessary for any field of modern science to make efficient progress. There is tremendous benefit to be derived if different research groups working on similar and complementary problems can exchange ideas and collaborate. Investigators free to follow where their research leads produce qualitatively new insights and new approaches to the underlying science. This type of scientific environment typically produces new techniques and new ideas on a rapid time scale. Such basic science is the foundation for applied science. These processes do not happen as frequently at large "technology centers," which must operate in a less flexible and more programmatic fashion.

In plasma science, the place for new and significant discovery is very frequently the laboratory. When there are significant new theoretical predictions, much of the value of these predictions is lost if they cannot be tested quantitatively by experiment. There is no adequate substitute for carefully planned and precisely controlled laboratory experiments. Many interesting and stimulating observations of plasmas can be made by spacecraft, but space experiments are not a substitute for the well-controlled and repeatable experiments that can be performed in the laboratory. The notion that computers can simulate plasmas so well that laboratory experiment can be replaced is also incorrect and is likely to remain so for the foreseeable future.

Laboratory experiments, theory and modeling, spacecraft and astrophysical observations, active space experiments, and experiments on fusion plasmas are synergistic. It is the healthy interplay among all these elements that will lead to a healthy plasma science. The field can "get along" for a while ignoring one element or the other, but it cannot continue for long in the unbalanced manner that has occurred in the last decade in the case of basic laboratory plasma experiments. Without a healthy underpinning of experimental laboratory work, the field of plasma science will not attract talented young scientists and is destined to become sterile and inefficient.

As a consequence, the highest priority of the panel is the establishment of a system of sustained support for modest-sized experimental efforts, sufficiently small and flexible that they can make rapid changes in their approach to a research problem, guided by the internal logic of the science and by new experimental and theoretical discoveries as they develop. As discussed in Chapter 4, fundamental aspects of plasma science crucial to fusion physics must be pursued in detail in the fusion-relevant geometries provided by large plasma devices and facilities. However, given the relatively high costs of operation of large facilities and the limited funds that one can expect for fundamental plasma experiments, the number of large devices not motivated by important applications such as space science or fusion is likely to remain small for the foreseeable future.

It is the conclusion of the panel that the type of sponsorship necessary for a revitalization of basic plasma experimental science is the support of at least 30 to 40 independent groups at a reasonable level for experimental research in a university, which is of the order of $200,000 to $400,000 per year. For example, funding at the $200,000 level would allow a program of two students, a postdoctoral researcher, and modest expenditures on equipment and supplies. Larger programs would require some technical support and additional personnel and equipment, as appropriate. It is important that such support be granted for periods of at least three years, given satisfactory performance, so that significant research goals can be set and accomplished. Large equipment purchases would have to be funded from separate equipment proposals. Given that the infrastructure for basic experimental facilities has declined so significantly in the past two decades, additional initial equipment purchases, where necessary, would typically range from $300,000 to $600,000 per program. Additional mechanisms that would allow for collaboration between groups on a rapid time scale, compared to the proposal cycle that now exists, would also be beneficial. Placing some resources at the discretion of program managers would be one way to accomplish this.

The increased support that the panel recommends for basic experimental research can be expected to serve an important educational function as well. It is generally recognized that small-scale experiments are an excellent setting in which to train students. The training of students, under the guidance of their supervisor, to make qualitative changes in an experiment or even in research direction as results unfold is invaluable in modern research and technological development. In addition, experimental plasma science students typically receive very thorough training in such important areas of modern technology as digital electronics, optics and computational hardware and software.

The following of the panel's general recommendations (see Executive Summary) are made to implement the revitalization of experimental plasma science described above:

1. To reinvigorate basic plasma science in the most efficient and cost-effective way, emphasis should be placed on university-scale research programs.
2. To ensure the continued availability of the basic knowledge that is needed for the development of applications, the National Science Foundation should provide increased support for basic plasma science.
3. To aid the development of fusion and other energy-related programs now supported by the Department of Energy, the Office of Basic Energy Sciences, with the cooperation of the Office of Fusion Energy, should provide increased support for basic experimental plasma science. Such emphasis would leverage the DOE's present investment in plasma science and would strengthen investigations in other energy-related areas of plasma science and technology.
4. Approximately $15 million per year for university-scale experiments should be provided, and continued in future years, to effectively redress the current lack of support for fundamental plasma science, which is a central concern of this report. Furthermore, individual-investigator and small-group research, including theory and modeling as well as experiments, needs special help, and small amounts of funding could be life-saving. Funding for these activities should come from existing programs that depend on plasma science. A reassessment of the relative allocation of funds between larger, focused research

programs and individual-investigator and small-group activities should be undertaken.

The panel recommends that the National Science Foundation increase its support for individual principal investigators conducting university-scale programs in basic research because this is most closely associated with NSF's mission. Increased support for basic research by the Department of Energy is recommended because DOE is charged with responsibility for both the magnetic and inertial confinement fusion programs, as well as a number of other energy-relevant programs that are critically dependent on the fundamental principles of modern plasma science.

9

Theoretical and Computational Plasma Physics

INTRODUCTION AND BACKGROUND

Plasma physics is the study of collective processes in many-body charged-particle systems. Like the fields of condensed matter physics and molecular biology, plasma physics is founded on well-known principles at the microscopic level. In the case of plasma physics, the description is based on the Liouville equation or kinetic equations for the electron and ion distribution functions in a multidimensional phase space and Maxwell's equations, whose sources are self-consistent moments of the distribution functions. The plasma state is distinguished by the existence of a vast number of collective motions over a very wide range of spatial and temporal scales. The interaction of these collective motions often leads to turbulence or coherent patterns and structures. Indeed, coherent patterns frequently may coexist with turbulence. A priori theoretical prediction of plasma behavior has enjoyed only limited success. Therefore, experiments are critical to the identification of fundamental processes in a plasma, such as the evolution of coherent structures arising from nonlinear interactions. These, in turn, form the intellectual building blocks for understanding the evolution of yet more complex processes.

The history of plasma science is as diverse as the subject itself. In Chapter 8 above, early work in laboratory plasma science is described, beginning with the work of Faraday in the 1830s on the chemical transformation of the elements and continuing with Langmuir's work on gas discharges in the 1920s and research on electron beams and beam-type microwave devices in the 1940s and 1950s. Within a decade of Langmuir's work, the discovery that radio waves reflect from the ionosphere established the existence of the space plasma that surrounds the Earth. A new era in plasma physics began with the international development of efforts to achieve controlled thermonuclear fusion in the 1950s and with the space program, which began with the launching of Sputnik in 1957. For the past 30 years, space, fusion, and the development of advanced weapons systems have been the main drivers for plasma science.

Early in the space and fusion programs, a rich variety of fundamental configurations and phenomena were investigated, but as a rule, nonlinear processes—although fascinating scientifically—proved to be a detriment to the achievement of fusion plasma conditions in the laboratory. As a consequence, fusion research evolved to focus on systems with the least complexity consistent with programmatic goals. Inertial fusion research evolved in directions that either minimized nonlinear laser-plasma interactions or optimized particle-beam drivers. Magnetic fusion research concentrated on the tokamak approach, the most stable axisymmetric confinement configuration. The principal difficulty encountered in fusion and in defense applications has been the inability to predict the nonlinear behavior of plasmas to an accuracy required by engineering considerations. A successful example of such a prediction is illustrated in Figure 9.1.

In the exploration of space plasmas, it was not possible to reduce the natural complexity of the magnetic field geometry through engineering design. Spacecraft data have identified many key nonlinear phenomena: collisionless shocks, bursty and steady magnetic reconnection, double layers, current sheets, dynamo generation of magnetic fields, and the overall structure of magnetospheric plasmas, which are high-mirror-ratio magnetic confinement configurations. Up until now, because spacecraft obtain local data, only the rudimentary aspects of these processes have been measured.

While the discoveries of plasma phenomena in the space environment are remarkably varied, their abstractions into basic plasma processes subject to investigation by computational simulation, laboratory experiments, and analytical theory have lagged because support, especially for laboratory experimentation, has "practically vanished" in the words of the Brinkman report, *Physics Through the 1990s*.[1] Notable exceptions exist, of course, and these are presented later in this chapter.

The next decade could promise a fundamental reversal of this paradigm, provided the resources for basic plasma experimentation described in Chapter 8 become available. One can anticipate that plasma phenomena discovered through spacecraft and astronomical observations, as well as fusion research, will play an important role in motivating laboratory experimentation. Moreover, the theo-

[1]National Research Council, *Plasmas and Fluids*, in the series *Physics Through the 1990s*, National Academy Press, Washington, D.C., 1986.

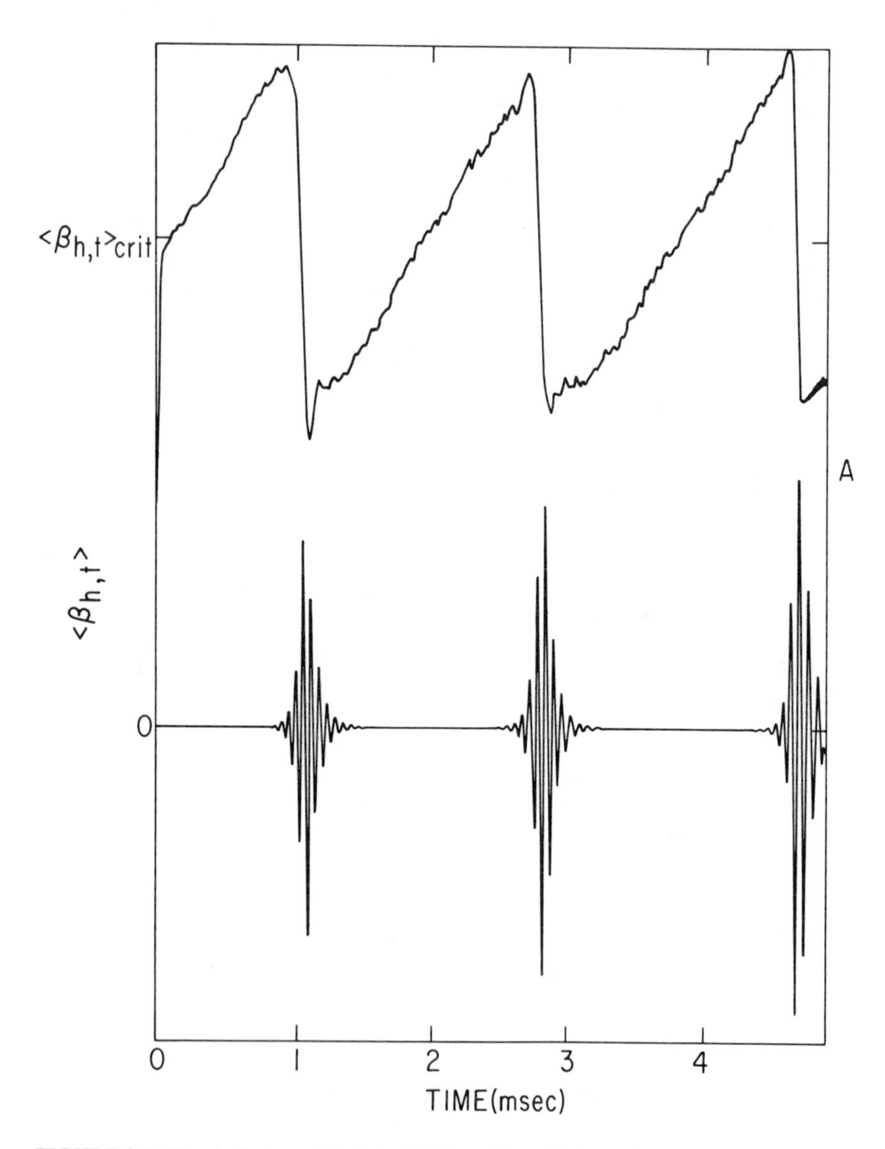

FIGURE 9.1 Theoretical model of the "fishbone" oscillations observed in tokamak plasmas. This figure shows that the high-frequency modulation of the magnetic field occurs in bursts (lower trace); it also shows the induced loss of high-energy particles during each burst, by the decrease in the normalized pressure of the energetic particles, β_h. (Reprinted, by permission, from L. Chen, R.B. White, and M.N. Rosenbluth, Physical Review Letters 52:1122, 1984. Copyright © 1984 by the American Physical Society.)

retical and simulation capabilities developed to understand this new generation of small- to intermediate-scale laboratory experiments will set standards for modeling space and astrophysical plasmas. Technological advances promise to create fundamentally new classes of plasma experiments and to enable new diagnostics. For example, as discussed in Chapter 8, our conceptual understanding of plasma dynamics will be enriched by visualization techniques only now becoming available for plasmas.

RECENT ADVANCES IN THEORETICAL AND COMPUTATIONAL PLASMA PHYSICS

Fusion, space exploration, and defense applications have been the engines of high national priority that have powered fundamental advances in plasma theory and computational plasma physics. In turn, the improved understanding of basic plasma processes has led to the seminal development of important new concepts and applications. Without attempting a complete delineation of significant achievements, in this section the panel highlights selected advances in analytical and computational plasma physics during the past decade that have resulted from the interaction of plasma theory with laboratory experiments and, to some extent, with space and astrophysical plasma measurements, because similar physics manifests itself in plasma systems of vastly different physical scales. At present, laboratory experimentation is dominated by research on magnetic and inertial fusion. Smaller experimental efforts can be found in active space experiments, nonneutral plasmas, coherent radiation generation, advanced accelerator concepts, and turbulent Q-machine plasmas. Additional advances in plasma theory and computations are incorporated in the chapters of Part II covering specific plasma topics.

Hamiltonian Transport

Apparently dissipative processes, such as particle diffusion, can occur in conservative Hamiltonian systems whenever chaos is present. In the 1980s, advances in understanding such transport were driven largely by anomalies observed in hot, effectively collisionless, magnetically confined plasmas, in which both the particle orbits and the magnetic field line trajectories obey Hamiltonian equations. A plausible contribution to anomalous loss is the effective diffusion induced by such chaotic behavior. Recent numerical and analytical studies have shown that Hamiltonian transport rates can depend sensitively on such unexpected structures as turnstiles, devil's staircases, and stochastic webs. Moreover, the ideas have been applied to estimating the loss of energetic charged particles from magnetically confined systems, reducing the necessity for elaborate and expensive numerical calculations based on guiding-center theory.

Coherent Structures and Self-Organization

The theoretical discovery of solitons, long-lived coherent solutions to certain nonlinear fluid equations, arose out of plasma physics research in the 1950s. In recent years, a more general class of nonlinear structures, including both solitons and less permanent, but still robust objects (solitary waves), has been found to play a significant role in plasma evolution. Thus, large-scale turbulence is often dominated by vortical structures, analogous to fluid vortices, but depending on the interaction of the plasma with electromagnetic fields. At smaller scales, such phase-space structures as clumps or holes can critically affect plasma dissipation. The past 10 years have seen significant progress in classifying, explaining, and assessing the importance of such phenomena. Their importance in laboratory plasma confinement, nonneutral plasmas, magnetosphere evolution, and solar physics is now firmly established, although much of the difficult nonlinear physics remains to be understood.

Strong Plasma Turbulence

The past decade has contributed to a greater, although still incomplete, understanding of plasma turbulence. Strong turbulence theory has been applied to many microinstabilities (drift waves and ion temperature gradient, trapped-particle, microtearing, and magnetic modes in tokamak plasmas). Resonance broadening has been addressed, and the direct interaction approximation (DIA), although still heuristic and difficult to implement numerically, has been extended to provide a general form that can be used for simpler transport models. Some understanding has been developed of turbulent cascades in plasmas, of nonlinear transport mechanisms, and of the coupling of heat and particle transport. Numerical studies of solar convection have greatly improved our understanding of turbulence in stars.

Gyrokinetics

During the past decade there has been a refinement of gyrokinetics, the approximate theory of the motion of charged particles in strong magnetic fields, with applications to stability theory and magnetohydrodynamics. Of particular note is the successful application of this description to the numerical simulation of a class of slow instabilities, the ion temperature gradient mode, resulting in mode spectra in excellent agreement with tokamak experiments. Also, a hybrid magnetohydrodynamic-gyrokinetic code has successfully simulated both fishbone and toroidal Alfvén eigenmodes, although greater computing power is needed to definitively study the latter.

Large-Orbit Effects on Plasma Stability

The influence of large-orbit particles in a plasma on low-frequency stability was computed by the Vlasov formalism in the form of a modified energy principle. The observed stability of field-reversed configurations has been attributed to this effect. A formal theory of interaction of a dilute species of energetic particles (described by the Vlasov equation) with magnetohydrodynamic Alfvén waves (described by fluid equations) has been developed and applied, with quantitative success, to tokamak plasmas and to the magnetosphere. The complex geometry of tokamaks, which is periodic the short-way-around the doughnut, altered the Alfvén wave propagation and attenuation bands as periodic media generally do. Almost-undamped Alfvén wave modes emerged that could be destabilized by energetic particles with velocities comparable to the Alfvén speed. This development forms the basis on which to expect challenging physics when thermonuclear reactions take place in magnetically confined plasmas.

Three-Dimensional Magnetohydrodynamics

Three-dimensional resistive magnetohydrodynamic simulations have successfully modeled turbulent generation of toroidal flux in force-free reversed-field pinch experiments. Three-dimensional resistive magnetohydrodynamics further gives a good account of magnetic reconnection in tokamaks and associated magnetic oscillations, including spontaneous formation of singular current sheets. However, a few troubling enigmas remain to be explored.

Numerical Simulation of Plasma Processes

The numerical study of plasmas has advanced markedly during the past decade, with applications to the ionosphere, the magnetosphere, solar flares, solar pulsations, stellar convection, nonlinear magnetohydrodynamics, gyrokinetics, and so on. (See Plate 8.) The progress has been due to a combination of improvements in algorithms and the advent of cheaper more powerful computers, both supercomputers and workstations, that provide great power, rapid turnaround, and networking at very modest cost. The computational discovery of nonlinear coherences that compensate for linear damping of microinstability modes in tokamaks calls into question the use of quasilinear correlation functions to estimate transport consequences of microinstabilities in tokamaks.

Nonlinear Laser-Plasma Interaction

Virtually all of the many instabilities driven by intense electromagnetic waves interacting with plasma were identified theoretically and studied in laser-plasma experiments during the past decade. Key nonlinear signatures predicted by numerical simulations and theory, such as the production of very energetic electrons, were confirmed. Various control techniques were also demonstrated, including collisional suppression and laser beam incoherence. Progress in this area had a major impact on research in inertial fusion, leading to the use of shorter-wavelength lasers.

Nonlinear Processes in Ionospheric Plasmas

The interaction of high-power radio-frequency waves with plasmas, in particular the ionosphere, has stimulated the theoretical development of a coupled ion acoustic-Langmuir wave turbulence model, generically known as the Zakharov equations. Computational studies of this model have identified spontaneous creation of cavitons—small-scale density structures that self-consistently trap Langmuir waves. Recently, fluid representations of collisionless damping have become available that will further increase the sophistication of the Zakharov approach.

Barium cloud releases in the ionosphere stimulated development of a new form of two-dimensional turbulence with key differences from two-dimensional hydrodynamic turbulence. Simulations based on these equations exhibited striking similarities to experimental releases. The equations were further applied to naturally occurring striations in the equatorial F-region and again enjoyed quantitative successes, especially with regard to the spectrum of turbulence.

The cross-magnetic-field current of the equatorial electrojet drives $\boldsymbol{E} \times \boldsymbol{B}$ turbulence in the equatorial region of the ionosphere. The nature of this low-frequency turbulence has been studied by radar backscatter diagnostics and in situ rocket campaigns. The plasma is weakly ionized so that the basic equations are well formulated and robust. The cascade theory of turbulent eddies, in the direct interaction approximation, predicts the nature of the nonlinear interactions and the line-width of the frequency spectrum, and is in accord with observations and numerical computations.

Collisional Relaxation of Nonneutral Plasmas

The consequences of binary collisions in nonneutral plasmas have been predicted to depend dramatically on magnetic field strength. In particular, when the duration of a collision, based on the distance of closest approach, exceeds the cyclotron period, the magnetic moment becomes an adiabatic invariant and the relaxation of perpendicular velocities becomes exponentially small. Quantitative experimental confirmation of an exponentially small equipartition rate between parallel and perpendicular temperatures has been demonstrated.

Free-Electron Lasers and High-Power Microwave Sources

Significant progress has been made in the fundamental nonlinear theory of high-power, coherent radiation generation in free-electron devices such as gyrotrons and free-electron lasers, particularly in areas related to nonlinear saturation mechanisms and efficiency optimization, mode selection and phase stability, and the effects of stochastic particle orbits. This has led to a new generation of laboratory microwave sources that have applications ranging from heating and noninductive current drive in fusion plasmas to communications and radar.

RESEARCH OPPORTUNITIES

This section identifies future research opportunities of fundamental importance in theoretical and computational plasma physics, including basic plasma theory and applications to laboratory plasmas, and space and astrophysical plasmas. Additional research opportunities in plasma theory and computations are incorporated in other chapters of this report.

Basic Plasma Theory and Applications to Laboratory Plasmas

The creative interplay between theoretical and computational studies and laboratory experimentation has been the classic engine that advances scientific understanding. In recent years, computations have begun to serve partially the role of experiments, with "simulation experiments" revealing unanticipated structures and coherences. The panel's vision of research frontiers in plasma physics during the next decade presumes that a program in basic laboratory experimentation will come into being, so that the range of topics investigated will be appreciably broader than those topics tied almost exclusively to fusion physics and defense applications. Of course, this will serve to extend and test current theoretical capabilities and to present qualitatively new challenges. Hence, the synergism among basic plasma theory, laboratory experiments, and space and astrophysical plasma observations promises to play an appreciably stronger role during the next decade.

Continuing progress is expected to be made in all of the areas identified earlier in the preceding section, "Recent Advances." In addition, the following topics, surely not comprehensive or mutually exclusive, represent research opportunities of high intellectual challenge in basic plasma theory and applications to laboratory plasmas.

Nonlinear Plasma Processes

Much of the progress in plasma physics during the past has been made by linear (small-signal) theory, which has provided conceptual and in many cases quantitative understanding. But nonlinear theory is of great intrinsic interest and is essential for the description of most important applications of plasma physics that involve magnetohydrodynamics, kinetic theory, turbulence, the interaction of charged particles with intense electromagnetic fields, and so on. Therefore, increased attention should be given to nonlinear theory aimed at the development of new analytical and numerical tools.

Numerical Simulation

A most promising area for the future is that of numerical simulation, driven by continuing dramatic advances in computational speed and computer organization, and decreases in the cost of hardware. These ongoing improvements in hardware, coupled with the parallel design of new and more efficient algorithms, should allow the solution of many of the nonlinear problems that currently defy direct analytical solution. Numerical computation offers the best hope of dealing meaningfully with the large problems in complex geometries that characterize so many of the significant applications of plasma physics. One anticipates the development of teams of computational specialists, theorists, experimentalists, and engineers, organized to optimize the solution of particular large technical problems. Training of students for this type of operation should be encouraged in universities.

Novel Analytical Techniques

The challenge of nonlinear theory suggests the adaptation or innovation of novel analytical techniques. The use of percolation theory for certain transport problems in plasmas appears to be promising. The development of modern statistical analyses, perhaps employing artificial intelligence (symbolic dynamics), may lead to greatly improved data analysis and new physical insights. The transfer from pure mathematics of well-developed areas such as wavelet theory, which are relatively unknown in physics and engineering, offers great promise.

Boundary Layers

Boundary layers are of great importance in plasmas. These occur in such diverse applications as the sheath region near the first wall of a fusion reactor, the region in a coronal hole where the solar wind is emitted as the system changes from collision-dominated to collision-free, and magnetic reconnection in plasmas of interest in space. They are often distinguished by the need for a full kinetic theory, and they will require a synthesis of analytical boundary layer techniques and advanced numerical methods.

Kinetic Theory

In the past, much theoretical work has been done employing fluid theories in circumstances where they are strictly not valid, because fluid descriptions are more tractable than kinetic ones. Such treatments are commonplace in astrophysics, space plasma physics, and long mean-free-path fusion physics. Advances driven by novel analytical approaches coupled with advances in computation are highly likely. The systematic exploitation of dimensionless small parameters to obtain reduced kinetic descriptions is one promising approach.

Stochastic Effects in Evolving Plasmas

In plasma physics, the conceptual advances of nonlinear dynamics and stochasticity theory have been applied to abstracted, simplified problems, capable of formulation in terms of Hamiltonian dynamics. Examples include criteria for stochastic magnetic field line "orbits" in tokamak and stellarator devices, nonlinear wave-particle interactions, and particle orbits near magnetic field nulls in the magnetosphere. During the next decade a key challenge for theoreticians will be to incorporate stochasticity self-consistently into evolving plasma systems. Magnetospheric magnetic reconnection serves as a useful example. How will stochastic orbits alter the evolution of magnetic fields and field nulls when the extent of the stochastic regions must be self-consistently determined in terms of the evolving field? Diagnostic and data processing advances that can identify stochasticity in plasma systems must be developed concurrently.

Alpha-Particle Effects in Magnetically Confined Plasmas

The level of sophistication of theoretical models and instrumentation techniques, and the high quality of plasma conditions achieved in laboratory experiments have progressed to the point that the deuterium-tritium experiments planned on the Tokamak Fusion Test Reactor (TFTR) and the Joint European Torus (JET) are expected to elucidate important physics issues regarding the influence of alpha particles on plasma stability and transport processes. There are two major phenomena associated with the appearance of significant densities of alpha particles in fusion devices. First, coherent plasma oscillations can be excited by resonant interaction with the alpha particles, either at low frequencies, corresponding to the toroidal precession rate, or at high frequencies, corresponding to the direct interaction of alpha particles with shear Alfvén waves. Such collective oscillations are driven unstable by the alpha-particle pressure gradient and result in the rapid expulsion of a significant number of high-energy particles. Several such modes have already been predicted theoretically and identified on existing devices, where they are excited by high-energy particles produced by injected beams or ion cyclotron heating. Second, high-energy particle orbits are modified by the presence of toroidal field ripple, and if sufficiently large, ripple can cause the orbits of trapped particles to become stochastic and to be lost from the plasma in a time that is short compared to the slowing-down time.

Concept Improvement

As the tokamak confinement approach continues to make significant technical progress toward achieving the conditions required for fusion power production, it is increasingly important to improve the tokamak concept, particularly in the long-pulse and steady-state regimes that can lead to compact designs for economical power production. This will require a significant theoretical and experimental effort in the study of advanced-tokamak regimes that optimize the bootstrap-current fraction produced by the collisional equilibration between trapped and passing particles, improve the efficiency of noninductive current drive, optimize the current profile and plasma shapes, and explore regimes of enhanced confinement and increased plasma beta. To ensure fully equilibrium (i.e., "steady-state") conditions, these plasma regimes must be studied for pulse lengths longer than the characteristic time scales of plasma processes and plasma-wall interactions. This will require steady-state power handling, particle exhaust, and impurity control by divertors at high power fluxes, as well as the development of advanced plasma fueling, current-drive, and control techniques. These are key features of the Tokamak Physics Experiment (TPX) planned for operation around the turn of the century.

Nonlinear Interaction of Intense Electromagnetic Waves with Plasmas

During the next decade, more quantitative models must be developed that describe the nonlinear interaction and competition of plasma instabilities driven by intense electromagnetic waves. This is a timely challenge for many reasons. In laser-plasma experiments, these instabilities are being characterized in greater detail, using increasingly sophisticated diagnostics. Nonlinear models based on the Zakharov equations are already illustrating the rich competition between different instabilities, and successful comparisons with experiments are being made. Finally, advances in computational physics and computers are allowing improved simulations, including, for example, meaningful three-dimensional simulations of laser-beam filamentation by both ponderomotive and thermal mechanisms.

More quantitative models of the nonlinear behavior of laser-plasma instabilities would allow optimized regimes of operation for inertial fusion and improved interpretation of data from ionospheric and space plasmas. Because these instabilities involve the most basic plasma waves, improved understanding of the nonlinear behavior would be a very significant contribution to plasma science, no doubt stimulating new advances and applications.

Current-Carrying Plasmas with Flow

One research opportunity in laboratory plasma physics, which is synergistic with space and astrophysical plasmas (see next section, "Space Plasmas"), is the investigation of current-carrying plasmas with flow. In these plasmas, global and/or turbulent flows are essential to the physics. By contrast, magnetic fusion plasmas are effectively stationary. Flowing plasmas present a new challenge to the experimentalist to design facilities in which the desired phenomena occur and to develop scaling arguments that laboratory plasmas are representative of the physics of space and astrophysical systems. To the theorist and applied mathematician, flowing plasmas are no less of a challenge because of the presence of several varieties of discontinuities, which must be understood as isolated, often collisionless processes and then incorporated self-consistently into the overall model just as hydrodynamicists incorporate shocks into supersonic flows. Discontinuities abound in plasmas. Even flows with velocities well below the Alfvén speed lead to singular current sheets in tokamaks and in the solar corona. And one need look no further than the solar photospheric magnetic field, which is concentrated into small regions of high intensity, to recognize that any theoretical understanding of dynamo generation of magnetic fields must accommodate an extraordinary degree of spatial intermittency. There has yet to be a successful laboratory demonstration of a hydromagnetic dynamo. With adequate support, visualization diagnostics can be implemented that promise to yield insights into magnetohydrodynamic flows comparable to those observed in hydrodynamic experiments.

Engineering Design Tools

The next decade should also see the development from the results of well-founded theories of simpler but robust tools for engineering design in the several areas of application of plasma physics, such as magnetic and inertial fusion, microwave devices, high-efficiency lamps, plasma processing, and particle accelerators.

Space Plasmas

In 1978, the National Research Council report *Space Plasma Physics: The Study of Solar System Plasmas,* prepared by a Space Science Board study committee headed by Stirling A. Colgate, strongly endorsed space plasma physics as "intrinsically an important branch of physics."[2] The Colgate report is widely considered to be the impetus for the present programmatic emphasis in the field.

---

[2]National Research Council, Space Science Board, *The Study of Solar System Plasmas*, National Academy Press, Washington, D.C., 1978.

Specifically, it identified six areas of research as important to develop the basic understanding of space plasmas and, therefore, of fundamental intellectual value to plasma physics. The areas are magnetic reconnection, turbulence, the behavior of large-scale flows, particle acceleration, plasma confinement and transport, and collisionless shocks. Substantial progress has been made in each area over the intervening 14 years. Examples are cited here and in Chapter 6, which is devoted specifically to space plasma physics. All of these areas offer significant research opportunities in theoretical and computational plasma physics during the next decade. Future research opportunities in space plasma theory are summarized below.

Magnetic Reconnection

Magnetic reconnection is a process by which stored magnetic energy can be converted explosively into plasma kinetic energy. It is invoked ubiquitously as a mechanism in space and astrophysical problems—in solar and stellar flares, which occur commonly in stressed bipolar magnetic regions; in flux transfer events, which intermittently erode the outer layer of Earth's day-side magnetic field and add the flux to a stressed antisunward magnetotail; and in the magnetospheric substorm, where the calamitous relaxation of the magnetotail is accompanied by pronounced auroral brightenings, enhanced electrical currents in the ionosphere, and the antisunward escape of a large blob of accelerated plasma. Observation of the reconnection process itself is fragmentary, and evidence is circumstantial. The mechanism occurs readily in models based on the magnetohydrodynamic (MHD) equations but is due to resistivity, externally postulated or spuriously generated by numerical discretization. Space plasmas are collisionless to a high degree, and reconnection is significant only when the resistivity due to classical two-body interactions is enhanced by anomalous collective processes. Candidate mechanisms such as the lower-hybrid-drift instability have been suggested. Further research needs to be carried out to investigate the microphysics and integrate the results into an MHD description of large-scale behavior. In turn, quantitative MHD predictions must benefit from the enhanced numerical capabilities of new computer architectures and algorithms that maximize their effectiveness.

Turbulence

Space plasmas are characterized by turbulence on all scale lengths. Hydrodynamic turbulence occurs in the convection zone of the Sun and, through coupling to rotation, may play an important role in the dynamo mechanism and global oscillation modes. Models that explore these processes on an elementary scale are being developed but are limited by numerical considerations. Magnetohydrodynamic turbulence is generated in the solar wind by the intersection of plasma streams emanating from different longitudes on the rotating Sun. This turbulence plays an important role in heating the solar wind. Indeed, the solar wind has become a fundamental medium for investigating MHD turbulence—its generation, evolution, and dissipation—with advanced statistical concepts. An unusual, microscale turbulence occurs when newly created plasma experiences the solar wind blowing across it, as is the case when cometary molecules are ionized. This turbulence is important in transferring momentum from the solar wind and can produce a deflection of the solar wind flow if the so-called mass loading is significant. Microturbulence plays many roles in magnetospheric plasma physics. Its importance in the magnetic reconnection process has already been mentioned. Another example, especially well studied from the standpoint of observations, modeling, and theory, is equatorial spread-F, a fluid-like turbulence in which flux tubes interchange in the low-latitude regions of Earth's collisional ionosphere.

Large-Scale Flows

The outward flow of the solar wind is one of the permanent features of space plasmas. Its speed varies by factors of two or three temporally and exhibits marked spatial variations, but the phenomenon is seen in every possible observation. Much less documented and understood is the coupling of such flow to planetary magnetospheres. Low-altitude observations at Earth indicate that the magnetospheric plasma is impelled into convection by coupling to the solar wind. The processes for this coupling are poorly understood: a steady magnetic reconnection may be the agent, or the magnetopause, the outer magnetospheric boundary, may experience Kelvin-Helmholtz instability so that the flow penetrates viscously. However, there is unanimity that the controlling processes take place in spatially localized boundary layers, with dimensions of the order of a few ion Larmor radii. The sense of the flow is antisunward at high latitudes and sunward at low latitudes, so that a circulation pattern is established. Because of the high electrical conductivity along magnetic flux tubes, it is expected that the entire magnetospheric plasma must participate in this convection. However, years of effort to measure steady magnetospheric convection far from Earth's surface largely have been inconclusive. The most recent data analysis suggests that the process may proceed in a bursty fashion. If so, large constraints would seemingly exist on the underlying physics: that it be initiated locally and hence probably involve microprocesses, and yet that it be transmitted globally on a rapid time scale.

Particle Acceleration

Because of their strong manifestations, energetic charged particles are central to astrophysics, and effort has focused there on acceleration processes that elevate their energy selectively from a thermal bath. Space plasmas are an accessible microcosm for examining some acceleration processes. Indeed, the observational origins of space plasma physics trace to cosmic-ray physics. Acceleration processes can be either systematic or random. An example of the former is the function of electric double layers that form at low altitudes along terrestrial magnetic field lines as a result of collective processes. Double layers have been created in laboratory devices and replicated in numerical simulations, but owing to their spatial extent, like so many phenomena they are difficult to identify unambiguously from space data. Much more widely invoked, because it is a process capable of raising particles to very high energies, is stochastic acceleration by random and repeated encounters with electromagnetic fluctuations. Outstanding success has been achieved in analyzing this mechanism in connection with shock-induced fluctuations. The treatment has been kinetic and self-consistent in the sense that the accelerated particles contribute to the energizing wave spectrum. An important next step is to assess the relative importance of such accelerated particles to the shock structure itself.

Plasma Confinement and Transport

Plasma confinement and transport are widely inclusive concepts. At the time of the Colgate study,[3] the primary reference was to the longevity of energetic, magnetically trapped particles, such as those populating the Earth's Van Allen belts. Like the solar wind, these are enduring features of the Earth's space environment, and analogous structures have been discovered in the environs of all the magnetized planets. The slow loss of Van Allen particles to Earth's atmosphere due to Coulomb collisions at very low altitudes and wave-particle scattering at higher altitudes is well understood. Of a much more speculative nature is the process that replenishes the belts so that they maintain their long-term existence. The issue is especially important at Jupiter, where MeV electrons are continuously losing energy due to synchrotron radiation. Present conjecture is that global-scale, low-frequency electromagnetic fluctuations, perhaps induced by solar wind buffeting of the magnetosphere or time-dependent atmospheric dynamo processes, allow particles to randomly traverse magnetic field lines to close-in distances, gaining energy in the process. Because of the global nature of the physics, it is difficult to verify this process observationally. What has been done is to create diffusion models using representative fluctuation spectra and to compare output particle distributions with observations—and this has been carried out with reasonable success. However, confinement and transport issues during the next decade will undoubtedly be far more expansive. To what extent are the outer reaches of the Earth's magnetosphere and the Sun's atmo-

---

[3]See footnote 2, p. 167.

sphere equilibrium structures? If the current direction of thinking is correct and gross plasma behavior is regulated by processes at a number of thin boundary layers, such as stream-interaction regions in the solar wind and Earth's plasma sheet boundary layer, what processes are responsible for maintaining the existence of such narrow structures?

Collisionless Shocks

The study of collisionless shocks is arguably the area in which the most significant advances have been achieved during the past decade and where the impact of space on basic plasma physics has been most profound. Success has been achieved as the result of a coordinated approach to the problem, using sets of complementary observations analyzed in the context of contemporary numerical models that integrate relevant microscopic theory. The Earth's bow shock is observable on every spacecraft orbit that reaches the solar wind. The International Sun-Earth Explorer (ISEE) mission is a cooperative venture between NASA and the European Space Agency (ESA). ISEE, with its coordinated pair of maneuverable spacecraft having ideally suited apogees, featured bow-shock physics as a prime scientific objective and has contributed immeasurably to this understanding. Bow shocks have been observed in association with every planet and comet. They are necessary hydromagnetic structures allowing the supersonic, super-Alfvénic solar wind to slow down and divert around an obstacle in its flow. Collisionless shocks are also common coronal and solar wind phenomena. There is rich variety to shocks, and this diversity is still incompletely investigated and understood. A prime determinant is the direction of the magnetic field in the incoming flow with respect to the normal to the discontinuity surface. If the angle is large, the shock is a perpendicular shock and generally laminar and quiescent. If the angle is small, the shock is a parallel one and exhibits a great deal of pulsation, structural disintegration, and reforming. Owing to the geometry, different azimuthal sectors of the same planetary bow shock can be quasiperpendicular, while other sectors are quasiparallel. In both instances there is a wealth of fine-scale plasma structure, both internal and external to the discontinuity.

Chaotic Effects

During the past decade, realization of the importance of chaos has emerged in all branches of science. This is true of space plasma physics. The focus of attention has been on chaotic particle trajectories, especially those occurring in the equatorial region of the magnetotail where the magnetic field is significantly weakened owing to distention. It is known that a mixture of chaotic and regular orbits exists, but the relative magnitude of their numbers is an open question. First results indicate that injection at the proper locations and ensuing chaos can lead to the formation of charged-particle beams, which are commonly observed. Particles on chaotic orbits also contribute significantly to the self-consistent current density responsible for the magnetic distention. If chaotic orbits are important, they must also certainly impact collective kinetic phenomena. Study of this subject is still in its infancy. It is quite plausible that similar physics takes place in the other boundary layers mentioned earlier in this section.

Summary

To summarize, space plasma physics has reached a mature phase. Space plasma physics has passed through the exploratory phase and is beginning an era of understanding. Future efforts will focus on the details of specific structures and processes. Theory and modeling at all levels have become a routine element of missions. As the morphology of and scientific basis for structures and processes become clearer, it is important that a basic understanding of the underlying plasma physics be pursued directly by relevant laboratory simulations, to test quantitatively and in detail the predictions of new theories as they become available.

CONCLUSIONS AND RECOMMENDATIONS

Plasma physics is the study of collective processes in many-body charged particle systems. The state of such a system departs considerably from the strictly thermodynamic equilibrium state. The understanding of collective processes is the central goal of theoretical plasma physics. There have been significant advances in theoretical and computational plasma physics during the past decade. Priority should be assigned to the research opportunities of high intellectual challenge identified earlier in this chapter. In the remainder of this section, the panel presents a succinct summary of principal findings and recommendations.

Fusion, space exploration, and defense applications have been the engines of high national priority that have powered fundamental advances in plasma theory and computational plasma physics. In turn, the improved understanding of basic plasma processes has led to the seminal development of important new concepts and applications.

The panel recommends that a vigorous program in plasma theory and modeling efforts in fusion and space exploration should be continued. This is essential for continued progress in the interpretation of experiments and the development of new concepts in these important national programs.

Support for individual university investigators to explore fundamental plasma theory and innovative concepts is at precariously low levels. This situation threatens the continued development and nurturing of the very intellectual foundations of modern plasma theory.

The panel recommends that the program of individual-investigator research in basic plasma theory should be reinvigorated to explore the broad range of intellectually challenging problems in stochastic effects, novel analytical techniques, nonlinear processes, and other areas, which are essential to the continued vitality of plasma theory as a scientific discipline.

Plasma theory is sufficiently advanced that a predictive capability exists for describing the properties of many static plasma configurations and simple wave-particle interactions. However, flowing, turbulent, and highly nonlinear saturation processes are at the forefront of analytical and computational capabilities.

The panel recommends that the design of experiments jointly by theoreticians and experimentalists to elucidate the conceptual foundations of nonlinear plasma physics should be encouraged.

The panel recommends that plasma theory should be encouraged that seeks to establish a commonality of physical processes and applied mathematical techniques across a wide range of realizations, from pellet compression in inertial fusion to plasma processes on astrophysical scales.

Advances in nonlinear plasma science in the past have relied heavily on the insights gained from numerical simulation. The panel envisions that future advances in theoretical plasma physics will have even greater reliance on numerical techniques and on the increased computational capability and visualization techniques available in present-day and future computer systems. A particular challenge is posed by discontinuities such as shocks, current sheets, and double layers.

The panel recommends that emphasis should be placed on ongoing programs in grand-challenge computations. Emphasis also should be placed on plasma computations investigating processes common to a wide range of scales.

The following of the panel's general recommendations (see Executive Summary) are made to improve the national effort in theoretical and computational plasma science:

1. To reinvigorate basic plasma science in the most efficient and cost-effective way, emphasis should be placed on university-scale research programs.

2. To ensure the continued availability of the basic knowledge that is needed for the development of applications, the National Science Foundation should provide increased support for basic plasma science.

3. Individual-investigator and small-group research, including theory and modeling as well as experiments, needs special help, and small amounts of funding could be life-saving. Funding for these activities should come from existing programs that depend on plasma science. A reassessment of the relative allocation of funds between larger, focused research programs and individual-investigator and small-group activities should be undertaken.

10

Education in Plasma Science

For plasma physics to make the contributions in the areas identified elsewhere in this report, there must be enough researchers and applied scientists knowledgeable in the plasma fields to enable those contributions to be made. To examine the demographics of the plasma physics field, data were obtained from the American Institute of Physics (AIP), the National Science Foundation, the National Research Council (NRC), and a survey of doctorate-granting universities.

The data indicate how many scientists were educated at the doctoral level in plasma physics. These scientists are not all of those now working in the field. The survey information also indicates how many new doctoral-level researchers will be entering the job market in the next five years. These, along with the current practitioners, must meet the challenges for plasma physics identified in the rest of this report.

DEGREE PRODUCTION AND EMPLOYMENT STATISTICS

From 1965 to 1991, 1539 doctorates were awarded in plasma physics. The average annual production in the 1970s was 72, with an upsurge between 1970 and 1972, peaking at 93 PhDs in 1972. The average annual production in the 1980s was 55. The number dropped to 42 in 1990 and rose to 58 in 1991.[1]

[1]National Science Foundation, *Science and Engineering Doctorates: 1960-1991*, NSF 93-301, Washington, D.C., 1993, Table 1.

Graduates in plasma physics have been primarily white male U.S. citizens: of the 1960-1991 total, nearly 97% were male and 74% were U.S. citizens. Of the 1976-1991 plasma physics doctorate recipients, 65% were white. In 1991-1992 there was a change in the nongender categories: plasma physics PhD recipients were 61% U.S. citizen and 53% white, but still 96% male.

The AIP provided data on employment based on a sample estimate of approximately one-tenth of all plasma physicists who are members of the AIP.[2] This information indicates that for PhD AIP members working full or part time in plasma physics in 1990, there is not one predominant category of employer. Four national laboratories, the university sector, and industry each account for about one-third of the positions:

University or university-affiliated research institute	34%
Federally Funded Research and Development Center (FFRDC)	34%
Industry	23%
Government	8%
Self-employed	1%

Most employees of FFRDCs were at Lawrence Livermore National Laboratory (LLNL), Princeton Plasma Physics Laboratory (PPPL), Los Alamos National Laboratory (LANL), and Naval Research Laboratory (NRL).

Plasma scientists associated with universities are often on the research staff of the university, not the teaching faculty:

Research staff (e.g., research scientist)	47%
Professor	33%
Associate professor	6%
Assistant professor	6%
Other/unknown	8%

Although the data do not indicate tenure-track versus non-tenure-track positions, the predominance of research staff positions suggests strongly that a large number of plasma physicists in university-associated positions are not on the tenure track.

Similar data for other fields in 1990 are given in Table 10.1. In none of these fields do physicists appear as likely to be in a non-tenure-track position as in plasma physics. This point probably is not missed by graduate students selecting a field.

---

[2]Information from AIP Statistics Division, included with letter from Jean M. Curtin, research associate, to John Ahearne, September 16, 1992.

TABLE 10.1 Employment Category in 1990 (percent of total) for University-affiliated Physicists in Selected Fields

	Nuclear Physics	Condensed Matter Physics	Atomic and Molecular Physics	Elementary Particles and Fields	Optics
Research staff	17	21	21	20	27
Professor	56	46	48	59	40
Associate professor	11	15	16	10	15
Assistant professor	9	17	11	10	17
Other/unknown	7	1	4	1	2

Source: Information from AIP Statistics Division, included with letter from Jean M. Curtin, research associate, to John Ahearne, February 9, 1993.

TABLE 10.2 Area of Employment in 1990 (percent of total) for Holders of PhDs in Selected Physics Fields

	Nuclear Physics	Condensed Matter Physics	Atomic and Molecular Physics	Elementary Particles and Fields	Optics
In field	29	45	28	40	58
Other physics	47	31	47	40	26
Engineering	7	10	9	8	6
Other/unknown	17	14	16	12	10

Source: Information from AIP Statistics Division, included with letter from Jean M. Curtin, research associate, to John Ahearne, February 9, 1993.

In 1990, about half of those who held a PhD in plasma physics were working primarily on plasmas; their subfields of employment were as follows:[3]

Plasma physics	51%
Other physics	30%
Engineering	12%
Other/unknown	7%

For comparison, Table 10.2 indicates areas of employment in 1990 for holders of PhDs in other physics fields. For all these degree fields, at least 75% of PhD

[3]Information from AIP Statistics Division, included with letter from Jean M. Curtin, research associate, to John Ahearne, February 9, 1993.

recipients were in some field of physics. However, only for plasma and optics were at least one-half the doctorate holders working in the same field as their doctorate.

Although many who were educated as plasma physicists have switched to other fields, crossover into plasma physics has also occurred. The AIP data indicate that many people working in plasma physics were educated in other fields. For those who indicated in 1990 that they were working full or part time in plasma physics, the predominant degree was as follows:[4]

Plasma physics	56%
Other physics	27%
Engineering	8%
Mathematics and statistics	1%
Unknown	8%

ESTIMATE OF FUTURE SUPPLY OF PLASMA PHYSICISTS

The NRC Doctorate Records Project identified 52 U.S. academic institutions that awarded at least one doctorate identified as in the field of plasma physics during the period from 1987 to 1991.[5] These 52 institutions awarded a total of 298 doctorates in plasma physics during this period. The data do not indicate from which department the degree was awarded.

Questionnaires were also sent to the chairs of physics departments (or other departments, if they had been identified as more appropriate). Responses were received from 40 departments, representing 38 institutions. The responding institutions produced 255 PhDs in 1987-1991 (86% of the total identified by the Doctorate Records Project). The departments estimated that during the next five years they would produce 332 to 340 PhDs in plasma physics—an increase of at least 11% over the previous five years. In the respondents' departments, in addition to 374 students in doctorate programs, there were 31 students in master's programs. Thus, if the previous supply of plasma physicists was enough to meet the needs of the field (implied by nearly one-half not working in plasma physics), unless there is a very large growth in demand the current estimated production rate should be more than adequate. It may be too high, which could be true of all physics, as indicated by the head of the AIP Education and Employment Statistics Division, Roman Czujko: "Results from AIP's most recent surveys indicate that the total number of projected vacancies in academia, government, and national laboratories combined is well under 50% of the total number of

[4]See footnote 3.

[5]Information included in letter from Lori Thurgood, research associate, NRC Office of Scientific and Engineering Personnel, to John Ahearne, October 13, 1992.

physics PhDs produced each year (approximately 1,260 in 1991). . . . There is no way that hiring in industrial settings will make up the difference completely."[6]

The fusion program is responsible for a large number of prospective plasma PhD graduates. Of the seven schools that reported that they expected to award at least 20 PhDs in plasma physics over the next five years, five have strong fusion programs, including the top four in expected graduates. These seven schools account for 57% of the total expected PhDs.

Of the 40 departments that responded to the questionnaire, 22 were physics departments. Others included space or astrophysics under various titles, applied physics, and several engineering departments. The number of "required" credit hours of course work for a plasma physics PhD averages 12.5 for the schools indicating a requirement; many recommend, but do not require, specific courses.

EDUCATING NON-PLASMA STUDENTS IN PLASMA PHYSICS

Of the 40 departments that responded to the questionnaire, 25 reported that they still have a plasma program. All 25 of these also offer courses for non-plasma science students. Usually this is a one- or two-semester course in plasma physics. Other courses offered include fusion, plasma transport, kinetic theory, and various space-related topics. The number of students taking these courses ranges from 1-2 to 20, with the usual number being 5 to 10. The bulk of these students are from physics or an engineering discipline.

GENERAL COMMENTS

Although degree production in plasma physics appears reasonably good, based on the number of expected PhDs, there are signs of erosion. Only 63% of the responding departments offer a major or a formal program in plasma physics, with 13 departments indicating they expected to award no doctorates in plasma physics in the next five years, though they had awarded a total of 27 during 1987-1991.

The following comments are typical of those provided in response to the question, If you no longer have a program to award a doctorate in plasma physics, why did the program end?

- "Lack of interest on the part of students. . . . Our one professor retired."
- "Lack of interest and lack of appropriate faculty."
- "For many years I did have a research program and funding, . . . and a number of PhD students did their theses in my lab. This research is no longer funded, and there is no one in the department now doing plasma work."

[6] *APS News*, Vol. 2, No. 2, Feb. 1993, p. 10.

- "Lack of faculty interest and student demand."
- "Plasma physicists left or retired; research interests in our department and university changed."

As we consider arguments for strengthening basic plasma sciences in universities, the following comment from one of the respondents offers some suggestions:

> Students are trained in fundamental and advanced plasma physics, whose theses are in such diverse topics as accelerator physics, solar physics, magnetospheric physics, and ionospheric physics. These are areas which are still well-funded by NSF and NASA. . . . A degree in "pure plasma physics" is not advantageous to students these days, although the training in plasma physics of students who receive their degrees in these other areas is practically indistinguishable from the training of a student who elsewhere might get a degree in plasma physics, per se. For example, we require competence . . . in electromagnetic theory, classical mechanics, and nonlinear dynamics, quantum mechanics, kinetic theory, and fluid mechanics, as well as in the usual plasma physics topics.

Plasma physics is a foundation for many areas of physics. A similar, but broader point was made recently by Donald Langenberg, president of the American Physical Society:[7]

> I'm afraid that we have managed to convince some of our most able young students that the only thing worth doing if you're a physicist is working at Fermilab, CERN, or in a university physics department. And if you don't, you're a failure in life. It just isn't so. There are few educational and training environments that better fit a young person for a very broad array of activities than physics. The key is not to cut back on the number of physicists, but to become much more flexible in our thinking about what physicists do.

The panel believes that plasma theory would be useful to many areas of physics and has three specific suggestions: (1) There is a need for short books or early chapters in larger books that would develop "plasma literacy" for non-plasma scientists. Many graduate students would be prepared to invest time to develop a basic level of plasma knowledge, but would do so only if less than a full course were available. (2) There is a need for senior-level, undergraduate texts on plasma science. (3) For more in-depth development, texts are needed that focus on disciplines.

Astrophysics serves as an illustrative example. The standard graduate curriculum in astrophysics contains graduate physics courses, such as quantum mechanics, electrodynamics, statistical mechanics, and classical mechanics. There also are standard astrophysics courses, such as stellar structure and evolution,

---

[7] *APS News*, Vol. 2, No. 2, Feb. 1993, p. 7.

stellar atmospheres and radiative transfer, interstellar medium, and galaxies and cosmology. At many universities, no courses in plasma physics are taught in the physics or astrophysics departments, and although plasma courses may be given in an engineering or applied science department, these often have too technological an orientation to attract astrophysics students. Depending on the inclination of the instructor, some plasma physics may be integrated into one or more of the astrophysics courses. However, very few astrophysics students receive much formal exposure to plasma physics and many astrophysicists view it as an arcane specialty.

Nevertheless, many astrophysicists might like to learn more plasma physics when motivated to do so by developments in their subject. For example, recent measurements of magnetic field strengths in dense, star-forming interstellar clouds have shown that the fields are large enough to strongly affect or even dominate the dynamics. This has spawned a real interest in MHD among interstellar medium researchers, and a number of people who ignored magnetic fields throughout most of their careers are now writing papers on them. Such people would benefit from a good, modern text on plasma physics, stressing astrophysically interesting applications and using astrophysically relevant parameters and boundary conditions. Such a book could consist of chapters contributed by experts, provided that a good editor and refereeing system kept the quality high, and could also be used for a graduate course or seminar.

RECOMMENDATIONS

There is an increasing emphasis on industry-university partnerships, with industry moving toward greater reliance on university research. Therefore, not only is the health of university plasma science important to the academic plasma community, but it is also important for industry to have a vibrant plasma effort in universities.

The panel recommends the following steps to improve education in plasma science:

- Both industry and academic members of the plasma community should use this report to support proposals to establish tenure-track positions in plasma science. Obtaining such positions will increase the likelihood of better scientists remaining in the plasma science field, and will attract higher-quality graduate students. Particular emphasis should be placed on establishing positions in physics departments.
- Industry and academic members of the plasma community should work with faculty and administrators to provide a course in basic plasma physics at the undergraduate senior level. This would be valuable for students going on into plasmas, fusion, astrophysics, electronics, and so on.
- To better prepare scientists and engineers for the many areas in which

plasma science is important, as demonstrated in this report, plasma researchers and teachers should develop (1) texts on plasma physics focused on other disciplines (e.g., astrophysics); (2) texts specifically suited for plasma subfields (e.g., low-temperature plasmas, plasma chemistry, and plasma processing); (3) chapters on plasma science for texts in other disciplines to develop "plasma literacy" in non-plasma scientists and engineers; and (4) undergraduate, senior-level texts in plasma science.

PART IV

❖

Conclusion

Plasma science is a diverse enterprise spanning many of the fields of science and technology. Plasmas of interest range over tens of orders of magnitude in temperature and density—from the tenuous plasmas of the ionosphere to the ultradense plasmas created in inertial confinement fusion, and from the cool, chemical plasmas used in the plasma processing of semiconductors to the thermonuclear plasmas created in magnetic confinement fusion devices. In this report, the panel has specifically assessed the diverse subfields of low-temperature plasmas, nonneutral plasmas, magnetic and inertial confinement fusion, beams, accelerators and coherent radiation sources, and space and astrophysical plasmas. Each subfield is making important contributions to our society, and much can be expected of each of these enterprises in the future.

In today's fiscally constrained environment, with major pressures to reduce federal spending, it is impractical to expect significant increases in funding. However, it is important for government decisionmakers, both in the executive branch and in Congress, to understand the value of plasma science and to recognize that some relatively minor steps can produce significant long-term gains for the United States. Plasma science is exactly the type of scientific field that a recent National Academy of Sciences report[1] stresses the United States should support:

[1]National Academy of Sciences, *Science, Technology, and the Federal Government: National Goals for a New Era*, National Academy Press, Washington, D.C., 1993, p. 20.

> The United States should maintain clear leadership in some major areas of science. . . . among the criteria . . . are the following: . . . the field affects other areas of science disproportionately and therefore has a multiplicative effect on other scientific advances . . . .

The great diversity of the subfields and applications of plasma science poses significant challenges to ensuring the coordination of activities, the funding of research and development, and even the recognition of the contributions that plasma science makes to our society. Given this great diversity, one of the panel's central recommendations is that there is a need for increased coordination of the support for plasma-related science and technology.

The central theme of this report is that while plasma science has contributed significantly to our society and will continue to do so, it is not adequately recognized as a scientific discipline. Consequently, the basic aspects of plasma science are not adequately supported, and this is threatening the fundamental health and efficiency of our entire effort in plasma science and technology. Although the applications of plasma science are in reasonably good health, the same cannot be said for the underlying, basic plasma science. One of the central conclusions of the Brinkman report, *Physics Through the 1990s*,[2] was that there was a critical need for increased support of basic plasma science. If anything, this is an even greater concern today, particularly in the area of basic plasma experimentation. The panel's conclusion is that structures for the adequate support of basic plasma science are absent across most of the subfields assessed. Thus one of the panel's central findings is that if the nation's investment in plasma science is to be effectively utilized, this deficiency must be remedied. Given the stringent budgetary considerations that can be expected for the foreseeable future, any recommendation must necessarily be modest in scope and motivated by the most pressing needs. With this in mind, one of the panel's central recommendations is to point out two agencies in which increased support of basic plasma science could be of enormous benefit to the entire field.

Increased support for basic experimental plasma science in the National Science Foundation (NSF) would have great impact on a wide variety of applications. This support for broad-based basic research is most in keeping with NSF's charter, and plasma experiment is singled out because it is identified as the most critical need.

The nation's largest investment in plasma science is in the area of magnetic confinement fusion research and development, sponsored by the Department of Energy (DOE). This program would benefit enormously from increased support of basic plasma science. In addition, the DOE sponsors support for many other

[2]National Research Council, *Plasmas and Fluids*, in the series *Physics Through the 1990s*, National Academy Press, Washington, D.C., 1986.

important energy-related applications of plasma science. Thus, the panel concludes that increased support for basic experimental plasma science by the Office of Basic Energy Sciences in DOE would provide a critical addition to the entire plasma science enterprise in the United States.

The panel concludes that the most critical need is in the area of basic plasma experimentation. In order to rectify this present lack of support, the panel recommends that approximately $15 million per year be provided, and continued in future years, for university-scale experiments.

With regard to the subfields and applications assessed, the panel found that, in general, larger programs had fared better than individual and small-group efforts. Typically, it is these smaller activities that have provided a disproportionately large share of the new concepts, new inventions, and new ideas. Thus, the panel recommends reassessment of the relative allocation between larger focused research programs and individual-investigator and small-group activities.

The panel identified two steps that could be taken by the plasma science community to improve the health of plasma science in the United States. The panel concludes that there is a need for the plasma science community to work for increased "plasma literacy" in our prospective scientists and engineers. The panel recommends that the community encourage courses in basic plasma science at the undergraduate level. The panel also recommends that the community work to make the case for tenure-track recognition of plasma science as an academic discipline.

Plasma science has made significant contributions to our society and can be expected to continue to do so. The panel concludes that the few changes described above, although small in comparison with the total support of plasma science and technology in the United States, could aid immeasurably in ensuring that the potential of plasma science to contribute to our society is fully exploited.

APPENDICES

A

❖

Federal Funding Data

As part of the development of this report, all federal government agencies that were known to fund plasma science were contacted and asked to provide funding information for FY 1989 to FY 1992. Appendix B contains the original letter of request in 1993 and the follow-up letter in 1994. Appendix C lists the agencies that were contacted.

Unfortunately, little detailed information was provided except by the military research offices, the Office of Naval Research (ONR) and the Air Force Office of Scientific Research (AFOSR). The poor response was not primarily due to lack of effort on the part of program officers, but rather to the lack of coordination and identification in agencies. This lends further support to the recommendations made in this report.

Illustrative is the effort at the National Science Foundation (NSF). There is no plasma branch at NSF, hence, no obvious coordinator for plasma science. One NSF staff member took the lead to act as an unofficial coordinator. He was able to pull together some information on funding, which illustrates many of the problems identified in this report.

In January 1993, plasma science and technology was identified as existing in seven divisions and three directorates at NSF, with a total funding of $15.4 million. To expand the assessment, a key-word search was made of all 1989 grants (the latest year for which the appropriate database is available). This led eventually to identifying all grants having at least a 5% component of plasma science or technology. These represented $17.5 million in 6 directorates and 19 divisions, and were included in 60 program elements. The largest category was space plasma with 46% of the grants, followed by plasma technology with 30%

TABLE A.1 Plasma Science Funding (current million dollars)—AFOSR, ONR, DOE

Agency	FY 1989	FY 1990	FY 1991	FY 1992	Change 1989–1992
AFOSR[a]	6.5	6.27	6.08	6.38	–1.8%
ONR[b]	~3.0	3.0	3.0	2.8	–7%
DOE-AEP[c]	3.2	2.2	2.3	1.2	–63%
DOE-ICF[d]	4.5	3.0	4.2	4.4	–2.2%
Total	~17.2	14.5	15.6	14.8	–12%

[a]Combines Life and Environmental Sciences with Physics and Electronics.
[b]Assuming 3.0 for 1989.
[c]Advanced Energy Projects, in BES.
[d]Estimate of that portion of the program going to basic plasma science.

TABLE A.2 Plasma Science Funding (current million dollars)—NASA Space Physics

Program	FY 1989	FY 1990	FY 1991	FY 1992	FY 1993	FY 1994	Change 1989–1994
SR&T	18.8	20.3	20.0	19.6	19.5	19.4	+3.2%
Data[a]	~ 10	12.7	12.9	25.7	8.0	6.9	approx. –30%

Note: SR&T = Supporting Research and Technology.
[a]The 1989 number is approximate. The large increase in 1992 combines data analysis with nonplasma instrument costs, cameras, cosmic-ray instruments, and so on, for the Pioneer and Voyager missions.

and basic plasma science with 22%. The reviewer's conclusion was that "support for plasma science and technology at NSF is *very* broad and *very* thin."[1]

With the above identified weaknesses, Tables A.1 and A.2 present a picture of the problems of plasma science funding. The programs listed in Table A.1, which fund small efforts in basic science, have not kept up with inflation, which totaled 13% from 1989 to 1992,[2] much less expanded to match the potential of

[1]"Plasma Science and Technology at NSF," prepared by Tim Eastman, NSF Atmospheric Sciences Division, May 10, 1993.
[2] Using the consumer price index.

TABLE A.3 DOE Funding for Magnetic Fusion (current million dollars)

FY 1989	FY 1990	FY 1991	FY 1992	FY 1993	FY 1994	Change 1989–1994
345	317	284	332	327	322	–6.7%

the field. If the amounts listed in Tables A.1 and A.2 are combined with the approximately $8 million for space plasma identified at NSF, these areas receive about $40 million per year.

The final area, magnetic fusion, is described in Table A.3. Although an order of magnitude greater than that for basic plasma science, the funding for magnetic fusion also has not kept up with inflation.

❖

Letters to Funding Agencies

NATIONAL RESEARCH COUNCIL

COMMISSION ON PHYSICAL SCIENCES, MATHEMATICS, AND APPLICATIONS

2101 Constitution Avenue Washington, D.C. 20418

BOARD ON PHYSICS AND ASTRONOMY

(202) 334-3520
FAX: (202) 334-2791
EMAIL: BPA@NAS.EDU

February 17, 1993

____
____
____

Dear Dr. ————:

The National Research Council (NRC) has formed the Panel on Opportunities in Plasma Science and Technology (OPST) to conduct an assessment of the field of plasma science. The Panel has been charged to assess the health of basic plasma science as a research enterprise, identify opportunities in plasma science, and make recommendations to federal agencies and to the plasma science community.

As we have begun to evaluate the potential for the future, we realized that it is necessary to have a reference frame in which to work—a not-uncommon feature in physics. The specific reference frame we lack is the funding for the areas associated with basic plasma physics in recent years. Therefore, we ask your help in assembling an accurate profile of federal funding for plasma science research. It would be helpful to get this over a period of several years, in particular, over the period fiscal years 1989 - 1992. Please specify if the dollars are current or constant dollars. If the information is available, a theory versus experiment breakdown would be useful, as would any other breakout.

Attachment 1 is a list of the areas into which our panel has separated basic plasma science. This may be useful in deciding what to include in any data you can give us. Attachment 2 is a list of federal plasma science program managers to whom this request was sent. If there are others that you know of within your agency that are not included on the list, we would appreciate your forwarding a copy of this request to them.

We appreciate that this is a difficult time to make such a request, since budgets for the next year are being revised and your time is stretched thin. However, any information you can provide will be most useful to the Panel if received by March 1, 1993. Please send it to the following address:

Dr. Ronald Taylor
Panel on Opportunities in Plasma Science and Technology
Board on Physics and Astronomy
2101 Constitution Ave
Washington, DC 20418

or fax the information to Dr. Ronald Taylor, 202-334-2791.

Sincerely,

John Ahearne
Clifford Surko
Co-Chairs, OPST

Enc.: Topics List; Distribution List
cc: Dr. Donald Shapero
Dr. Ronald Taylor

TOPICS LIST

- Beams, accelerators, and coherent radiation sources
- Nonneutral plasmas
- Basic plasma science in magnetic and inertial fusion plasmas
- Space plasma physics
- Astrophysics
- Low temperature plasmas
- Fundamental plasma experiments
- Theoretical and computational plasma physics

NATIONAL RESEARCH COUNCIL
COMMISSION ON PHYSICAL SCIENCES, MATHEMATICS, AND APPLICATIONS
2101 Constitution Avenue Washington, D.C. 20418

BOARD ON PHYSICS AND ASTRONOMY

(202) 334-3520
FAX: (202) 334-2791
EMAIL: BPA@NAS.EDU

August 12, 1994

____
____
____

Dear Dr. ——————:

We are writing to you as Co-Chairs of the Panel on Opportunities in Plasma Science and Technology (OPST) to ask your help in assembling financial information for a report that the panel has prepared. This letter is actually a followup to a letter that we sent to you on February 17, 1993. Some background information and our specific request are outlined below.

The OPST, under the auspices of the Board on Physics and Astronomy (BPA) of the National Research Council (NRC) has been conducting an assessment of the field of plasma science. The report is nearly complete and is currently under review at the NRC. The original charge to the panel was to assess the health of basic plasma science as a research enterprise, identify research opportunities in plasma science, and make recommendations to federal agencies and to the plasma science community. In addressing this charge, the panel assembled information on federal funding for basic plasma science. Your response to our February 17, 1993 letter was used for that purpose.

Several attachments are included for your reference. Attachment 1 is a copy of the February 17, 1993 letter that was sent to you requesting funding information on basic plasma science. Attachment 2 is a copy of your response. Finally, Attachment 3 is a copy of the section of our draft report that summarizes the information we were supplied last year. Our request is the following:

(1) We would like to update the information that you provided last year. Please review your response to us and return funding figures for FY93 and FY94 (if possible).

(2) Our original request asked for funding information associated with basic plasma science. We would like to compare these figures with the total funding for plasma science-based research. It would be most helpful if you could provide us with those figures for FY89 - FY94.

(3) Finally, please review Attachment 3 and give us any comments. We would like this section of the report not only to reflect the panel's information on the funding situation in basic plasma science, but also the funding situation of the field. Therefore, we are interested in your perspective.

We appreciate that this is a difficult time to make such a request and thank you for your time and effort. Any information you can provide will be most useful to the panel if received by September 1, 1994. Please send it to the following address:

Panel on Opportunities in Plasma Science and Technology
Board on Physics and Astronomy
2101 Constitution Ave
Washington, DC 20418

or fax the information to Natasha Casey, 202-334-2791. Ms. Casey, from the BPA office, will call you in several days to confirm that you received this request and offer any assistance that you might need in responding to this request.

Sincerely,

John Ahearne
Clifford Surko
Co-Chairs, OPST

Enc.: Attachments
cc: Dr. Donald Shapero
Dr. Ronald Taylor

List of Agencies Contacted

National Science Foundation
- Physics Division
- Astronomical Sciences Division
- Atmospheric Sciences Division
- Electrical and Communications Systems Division (Quantum Electronics Waves and Beams Program)
- Office of Polar Programs (Polar Aeronomy and Astrophysics Program)

National Aeronautics and Space Administration
- Space Physics Division
- Astrophysics Division
- Solar System Exploration Division (Planetary Science Branch)
- Goddard Space Flight Center

Department of Energy
- Office of Basic Energy Sciences
- Office of Fusion Energy
- Office of Defense Programs (Office of Inertial Fusion)

Office of Naval Research

Army Research Office

Air Force Office of Scientific Research

Air Force Cambridge Research Laboratory (Ionospheric Physics Laboratory)

NOTE: In some cases letters were sent to multiple offices or programs within those listed above.